Ion Exchange Resins and Adsorbents in Chemical Processing

Ion Exchange Resins and Adsorbents in Chemical Processing

E.J.Zaganiaris

consultant

Second Edition

OF THE SAME AUTHOR:

Ion Exchange Resins in Uranium Hydrometallurgy, Books on Demand, 2009

Ion Exchange Resins and Synthetic Adsorbents in Food Processing, Books on Demand, 2011

Published by:
Books on Demand GmbH,
12/14 rond point des Champs Elysées
75008 Paris, France

Printed by:
Books on Demand GmbH, Norderstedt, Germany

© Copyright 2016 : Emmanuel Zaganiaris
ISBN: 978-2-322-112913
Dépôt légal: août 2016

To the memory of my wife

*for her support, encouragement and love
all these years*

Preface

Ion exchange is an established unit operation in a variety of industries such as industrial water treatment, food processing, potable water purification, pharmaceutical processing, catalysis and chemical processing.
Applications of ion exchangers in chemical processing deal in principle with manufacturing processes of chemicals, which includes raw materials purifications, process streams purification and recycling, removal of impurities in order to upgrade the final product, recovery of valuable products and metals from industrial liquids or plating baths as well as wastes treatment. Both, aqueous and non-aqueous media may be involved and gas streams can be treated as well.

The use of ion exchange in chemical processing is characterized by the great variety of compounds, both organic and inorganic, operations as well as mechanisms of "ion exchange" that are involved. When metals are to be removed or recovered, formation of complexes plays an important role. Some separation mechanisms, such as acid retardation, ion retardation and ion exclusion present particular interest because water is used as regenerant instead of chemicals. Many of the separation processes described in the text have been developed based on complexing resins, on their different selectivities for various elements or on selective elution. It is interesting to note that the number of chelating resins commercially available has increased in the recent years.

In this book, various examples are discussed on the use of ion exchange in chemical processing, mainly in aqueous systems but also in non-aqueous systems and in gas streams. The theory behind these examples is also briefly discussed in order to make the subjects better understood. It is noted nevertheless that the description of various processes given in this book are given as a suggestion for trying in a laboratory and using relevant patents with a license from the owner of the patent.

As it was the case in the previous books of this series, this one is also based on my experience while working in the ion exchange department of Rohm and Haas Company and from the numerous visits to companies or institutes while in the employ of Rohm and Haas. It is addressed to people working in the chemical processing industries for whom ion exchange is not their principal expertise.

Emmanuel J. Zaganiaris

Note on the second edition

This second edition includes many additions to the first one some of which have come up from discussions with the Dow Technical Service and Development group. I would like to thank more particularly Dr Klaus-Dieter Topp and Dr María de los Ángeles Pérez Maciá for discussions and for reviewing parts of the book.

E.J.Z

Acknowledgements

Acknowledgement is made here to Craig Brown and to George Di Falco of Eco-Tec Inc., and to Klaus-Dieter Topp of The Dow Chemical Company for kindly supplying useful information and other material and permission to use them in this book.

ABBREVIATIONS

AMD	acid mine drainage
AMP	aminomethylphosphonic
AOX	absorbable organic halogens
APM	anion permeable membrane
APU®	Acid Purification Unit
AVR	Acidify, Volatilize, Reneutralize
BPM	bipolar membrane
BV	Bed Volume
CEM	cation exchange membrane
CMP	chemical mechanical polishing
CPM	cation permeable membrane
D2EHPA	di-2-ethyl hexylphosphoric acid
DCP	dicalcium phosphate
DVB	divinylbenzene
EDI	electrodeionization
HMS	harmonic mean size
HSAB	hard and soft acids and bases
IDA	iminodiacetic
IPG	industrial pure grade
IX	Ion exchange
IER	Ion exchange resin
L_R	Liter of resin
MHC	moisture holding capacity
MP	macroporous
MR	macroreticular
MRT	Molecular Recognition Technology

PCB	Printed Circuit Board
PEI	Polyethyleneimine
PGM	Platinum Group Metals
PLS	Pregnant Leach Solution
PNC	Photonitrozation of cyclohexane
RH	Relative humidity
RIL	Resin-in-leach
RIP	Resin-in-pulp
SAC	strong acid cation
SAD	strong-acid dissociables
SBA	strong base anion
SIR	solvent impregnated resins
SSU®	Salt Separation Unit
SX	solvent extraction
VCM	vinyl chloride monomer
Vol cap	volume capacity
VOC	Volatile Organic Compounds
WAC	Weak acid cation
WAD	weak-acid dissociables
WBA	Weak base anion
Wt cap	weight capacity

CONTENTS

Preface
Acknowledgments
Abbreviations

1. Introduction	19
2. Chemical processing and ion exchange	21
Coordination complexes of metals and metalloids	23
Ion exchange resins with ligand functional groups	37
Weak base anion exchange resins	37
Iminodiacetic resins	38
Aminomethylphosphonic resins	44
Other phosphorus containing resins	46
Picolylamine resins	48
Thiol resins	52
Thiourea resins	54
Thiouronium resins	54
Dithiocarbamate resins	56
Polyamine resins	56
Amidoxime resins	58
Pyridine resins	58
Methylglucamine resins	59
Carboxylic resins	60
Special types of resins	
Amine-borane resins	61
Molecular Recognition Technology	61
Solvent impregnated resins (SIR)	62

- Separation processes without chemical regeneration ... 64
  - The Donnan equilibrium 64
  - Ion exclusion 69
  - Acid retardation 70
  - Ion retardation 73
  - Parametric pumping 75
  - Thermal regeneration 76
  - Electrodeionization 77
  - Water regeneration 78
- Some aspects of ion exchange equilibrum........... 79
3. Metals removal or recovery with IX 91
  - Metal processing industries 92
    - Pickling liquors purification 95
      - HCl 95
      - H_2SO_4 99
      - Mixed HNO_3/HF 99
    - Chemical passivation 102
    - Aluminum anodizing 104
  - Metal plating 107
    - Electroplating 107
    - Electroless plating 110
    - Chromium plating 111
    - Printed Circuit Board rinse waters 119
    - Nickel recovery from plating rinse waters ... 122
    - Copper recovery from plating rinse waters ... 123
    - Cadmium electroplating 124
    - Galvanization 124
    - Electroles copper and nickel recovery......... 127
    - Gold recovery 128
    - Silver recovery 130
    - Platinum Group Metals 130
    - Organics removal from rinse waters 132

    Waste treatment from metal plating 132
    Summary of ion exchange systems 135
    Copper and vanadium recovery from adipic acid…..137
    Zinc recovery from viscose rayon spinning effluents. 138
    Copper and ammonium recovery from cuprammonium rayon spinning effluents………………………...139
4. Hydrometallurgical applications 141
    Gold recovery 142
        Cyanides removal from wastes 155
    Platinum Group Metals 158
    Base metals 161
    Group IV: zirconium, hafnium. Group V: vanadium, niobium, tantalum. Group VI: molybdenum, tungsten. Group VII: rhenium……………………. 168
    Rare earths ………………………………………. 175
    Gallium …………………………………………… 179
    Acid mine drainage ……………………………… 180
    Zinc from geothermal brine ……………………… 182
5. Purification of chemicals and process streams ……… 185
    Brine purification in the chloralkali industry ……. 185
        Hardness removal …………………………. 190
        Mercury removal …………………………… 195
        Aluminum and silica removal ……………... 196
        Nickel removal …………………………….. 197
        Iodide removal …………………………….. 197
        Sulfates removal ……………………………. 199
        Lithium brines ……………………………… 201
        KCl brines ………………………………….. 202
    Boron removal from $MgCl_2$ brines ……………….. 202
    H_2O_2 purification …………………………………… 205
    Dimethylformamide purification …………………… 209
    Glycols purification ……………………………….. 210

  Caprolactam purification 212
  Formaldehyde deacidification 215
  HCl purification 216
  H_3PO_4 purification 218
  Caustic purification 221
  Photographic baths and wastes purification 223
6. Chemical synthesis 225
  Fertilizers 225
  Hydroponics 229
  Silica sols 229
  Butanol recovery from fermentation broths.......... 232
7. Waste waters purification 233
  Heavy metals removal from wastes 233
  Fluorides removal 236
  Arsenic removal.. 237
  Incinerators wastes treatment 238
  NH_4NO_3 recovery from wastes of fertilizer plants .. 240
  Organics removal from wastes 241
8. Air purification 245
9. Non-aqueous solutions 249
  Phenol deacidification 252
  Methanol purification 254
  Mercury removal from hydrocarbons 255
  Biodiesel purification 256

References 259
Subject index 289

1. Introduction

In chemical processing, ion exchange resins (IER) are used mainly:
* in purifying a chemical compound by removing impurities
* in removing a chemical from a solution and subsequently recovering it by elution
* in the synthesis of a product, (where the resin acts not as catalyst but as a chemical)
* in purifying and recycling process liquors and
* in treating waste solutions.

In view of the fact that under "chemical" a great variety of inorganic and organic compounds are included, the various chemical processing applications of IER are very versatile and involve different mechanisms of "ion exchange". Thus, in addition to the conventional ion exchange and adsorption mechanisms, we may have:

Complex exchange, where the element to be removed forms complexes with ligands found in the external solution. The formed complex is removed by exchange with the counterion on the resin. It may be that the same complex is also formed on the functional group of the resin when the ligand is the counter-ion fixed on the functional group.

Complex formation where the functional group of the resin is a ligand with which the ion to be removed from the solution forms complexes. It can be that there are two ligands involved, one of which is chemically bound on the resin while the other is found in the solution. In that case the removal of the ion from the solution depends on the relative stability of the two complexes involved.

Another mechanism is *ligand exchange* where the element or molecule to be removed from solution is a ligand while the counter-ion of the resin is a complexing metal. The removal of the ligand found in solution takes place by exchanging with the ligands of the metal fixed on the functional group of the resin.

Ion exclusion, acid retardation and ion retardation are three mechanisms which were developed by The Dow Chemical Company in the fifties-early sixties (Bauman 1954; Dow Chemical 1960; Hatch and Dillon 1963). With these techniques, one can separate ionic from non-ionic species, strong acids from salts or different salts from each other. The advantage of these mechanisms is that elution is achieved with water. *Parametric pumping* is another interesting devise that does not use chemicals for separation processes using ion exchange resins. *Thermal regeneration and electrochemical regeneration* are techniques that do not use chemicals for regeneration of the resins and can also be used in chemical processing applications.

An important part of chemical processing involves metals. In the following chapter it is discussed the various ion exchange mechanisms involved in metal processing with ion exchange, as well as the various types of complexing ion exchange resins developed for these applications. In addition, the above mentioned special separation processes are briefly discussed along with some basic ion exchange principles which are directly involved in metal removal from acidic solutions.

2. Chemical processing and ion exchange

The various applications of ion exchange (IX) in chemical processing mentioned above can be classified in the following operations:
- Removal of non-ionic compounds from solutions
- Removal of simple ionic compounds
- Removal of metals by formation of complexes

In the first case, the more frequent applications concern organic compounds and adsorption is the main mechanism involved. Synthetic adsorbents are regenerable, that is they are used in a cyclic way.

In the second case, ion exchange resins are used to remove simple inorganic or organic electrolytes. Strong acid cation (SAC), weak acid cation (WAC), strong base anion (SBA) or weak base anion (WBA) exchangers, gel or macroreticular (MR) types as well as styrenic, acrylic or formophenolic type resins can be used according to the requirements based on the operating conditions. In some cases, the ion exchange resins have a special chemical structure which allows the resin to develop high selectivities for certain electrolytes. Examples of these selective resins include the high selectivity of strong base anion exchange resins for monovalent over polyvalent anions. This selectivity is achieved with SBA resins having trialkylamine functional groups with more than two carbons, for example triethylamine, tripropyl, etc. the obtained high selectivity for monovalent anions is due to a steric effect since monovalent anions can get easier to the trialkylammonium group than polyvalent ones. Thus, there exist selective resins for nitrates or for perchlorates over sulfates and other polyvalent anions and for gold cyanides over polyvalent metal cyanides (see chapter 4). For perchlorates and pertechnetates, a bifunctional SBA resin has been developed at the Oak Ridge National Laboratory (Gu *et al*, 1999; Alexandratos *et al*, 2000) which increases even further the selectivity of monofunctional SBA exchangers. It was found in fact that as the number of carbon atoms in the alkyl group increases, the selectivity drops which was explained by poor kinetics and steric inhibition. By combining two different functional groups, like triethyl/trihexylamine groups, on the same resin it was found that selectivity increases and exchange kinetics improve. In the third case, metal (or metalloid) complexes are involved and in addition to conventional ion exchange resins, special resins, sometimes called selective or chelating resins, are used.

From these operations, the first two involve adsorption or ion exchange mechanisms with conventional ion exchange resins and adsorbents. An ion loaded on the resin is exchanged for another ion of the same sign found in solution. The exchange reaction is reversible and an equilibrium is established. In removal of metals on the other hand, a simple exchange of a metal cation in solution with a counter-ion loaded on the resin is frequently not practical. A conventional SAC or WAC exchange resin is not selective enough for metals and is loaded with all cations present in solution. In addition, a simple equilibrium involving the cations in solution is disturbed by the formation of metal complexes. The selectivity sequence that exists with conventional IER and simple ions is no longer valid and special resins are often required. In what follows, a brief review of the formation of metal complexes is presented and the special, chelating, resins developed are described.

A special case represents interactions between metals and synthetic adsorbents bearing no functional groups. The first is cation-p interaction. This results from electrostatic interaction between the cation and the π-face of the aromatic ring of an aromatic adsorbent. The other interaction is illustrated with certain metals in acidic media that can be adsorbed on acrylic adsorbents. The mechanism is possibly the interaction of the chlorocomplexes with the oxygen atoms of the acrylic adsorbent.

Coordination complexes of metals and metalloids

Metals represent the major part of the periodic table. Looking at the periodic table of page 25, the elements B, Si, Ge, As, Sb, Te, Po, colored grey, are called metalloids, the elements on the right of the metalloids are the nonmetals while the elements on the left are the metals.

A metal ion can form coordination complexes with neutral molecules or anions, called ligands, via interactions between lone pairs of electrons of the outer energy level of the ligand and the s, p and d orbitals of the metal. The ligand is a Lewis base (donates electrons) and the metal is a Lewis acid (accepts electrons). The bond between the ligand and the metal, where the ligand donates the electron pair, is called coordination bond, as opposed to a covalent bond where each atom provides one electron. Generally speaking, the strength of a coordination bond is in between that of a covalent bond and an electrostatic bond. It is this formation of complexes that allows the selective removal and eventually subsequent recovery of metal ions from solutions containing high concentration of other cations with complexing resins, as it will be discussed below.

Examples of common ligands are: H_2O, NH_3, Cl^-, OH^-, CN^-, represented with the dot Lewis structure by:

$$H:\overset{..}{\underset{..}{O}}:\quad H:\overset{H}{\underset{H}{\overset{..}{N}}}:\quad :\overset{..}{\underset{..}{Cl}}:^-\quad H:\overset{..}{\underset{..}{O}}:^-\quad :C:::N:^-$$

Take as example the $[Al(H_2O)_6]^{3+}$ complex. Aluminum has the electronic structure $1s^2 2s^2 p^6 3s^2 p^1$. In the Al^{3+} state it has the electronic structure $1s^2 2s^2 p^6$. Al^{3+} uses the 3s and 3p orbitals and two of the 3d orbitals to produce 6 orbitals with the same energy where it accepts six electron pairs of six water molecules:

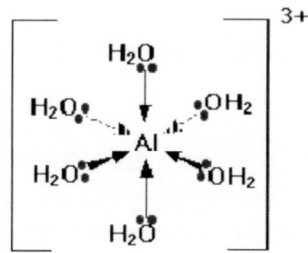

In this example, the H_2O molecule uses one electron pair to form a bond with Al^{3+}.

24

H 1s¹																	He 1s²
Li [He] 2s¹	Be [He] 2s²											B [He] 2s²p¹	C [He] 2s²p²	N [He] 2s²p³	O [He] 2s²p⁴	F [He] 2s²p⁵	Ne [He] 2s²p⁶
Na [Ne] 3s¹	Mg [Ne] 3s²											Al [Ne] 3s²p¹	Si [Ne] 3s²p²	P [Ne] 3s²p³	S [Ne] 3s²p⁴	Cl [Ne] 3s²p⁵	Ar [Ne] 3s²p⁶
K [Ar] 4s¹	Ca [Ar] 4s²	Sc [Ar] 3d¹4s²	Ti [Ar] 3d²4s²	V [Ar] 3d³4s²	Cr [Ar] 3d⁵4s¹	Mn [Ar] 3d⁵4s²	Fe [Ar] 3d⁶4s²	Co [Ar] 3d⁷4s²	Ni [Ar] 3d⁸4s²	Cu [Ar] 3d¹⁰4s¹	Zn [Ar] 3d¹⁰4s²	Ga [Ar] 3d¹⁰4s²p¹	Ge [Ar] 3d¹⁰4s²p²	As [Ar] 3d¹⁰4s²p³	Se [Ar] 3d¹⁰4s²p⁴	Br [Ar] 3d¹⁰4s²p⁵	Kr [Ar] 3d¹⁰4s²p⁶
Rb [Kr] 5s¹	Sr [Kr] 5s²	Y [Kr] 4d¹5s²	Zr [Kr] 4d²5s²	Nb [Kr] 4d⁴5s¹	Mo [Kr] 4d⁵5s¹	Tc [Kr] 4d⁵5s²	Ru [Kr] 4d⁷5s¹	Rh [Kr] 4d⁸5s¹	Pd [Kr] 4d¹⁰	Ag [Kr] 4d¹⁰5s¹	Cd [Kr] 4d¹⁰5s²	In [Kr] 4d¹⁰5s²p¹	Sn [Kr] 4d¹⁰5s²p²	Sb [Kr] 4d¹⁰5s²p³	Te [Kr] 4d¹⁰5s²p⁴	I [Kr] 4d¹⁰5s²p⁵	Xe [Kr] 4d¹⁰5s²p⁶
Cs [Xe] 6s¹	Ba [Xe] 6s²	*La [Xe] 5d¹6s²	Hf [Xe] 4f¹⁴5d²6s²	Ta [Xe] 4f¹⁴5d³6s²	W [Xe] 4f¹⁴5d⁴6s²	Re [Xe] 4f¹⁴5d⁵6s²	Os [Xe] 4f¹⁴5d⁶6s²	Ir [Xe] 4f¹⁴5d⁷6s²	Pt [Xe] 4f¹⁴5d⁹6s¹	Au [Xe] 4f¹⁴5d¹⁰6s¹	Hg [Xe] 4f¹⁴5d¹⁰6s²	Tl [Xe] 5d¹⁰4f¹⁴6s²p¹	Pb [Xe] 5d¹⁰4f¹⁴6s²p²	Bi [Xe] 5d¹⁰4f¹⁴6s²p³	Po [Xe] 5d¹⁰4f¹⁴6s²p⁴	At [Xe] 5d¹⁰4f¹⁴6s²p⁵	Rn [Xe] 5d¹⁰4f¹⁴6s²p⁶
Fr [Rn] 7s¹	Ra [Rn] 7s²	**Ac [Rn] 6d¹7s²															

For most ligands, the coordinating atom (the atom of the ligand that donates its electron pair to the metal) comes from groups V (N, P), VI (O, S) and VII (F^-, Cl^-, Br^-, I^-) of the periodic table.

As mentioned, the formation of a complex is the result of an interaction of a Lewis acid (the metal, electron acceptor) with a Lewis base (the ligand, electron donor). These acids and bases can be divided into hard acids, hard bases, soft acids and soft bases, according to the HSAB theory (Hard and Soft Acids and Bases) (Pearson, 1963). Hard acids are metals with high charge to radius ratio (Q/r) with low electronegativity and non-polarizable. Soft acids are the contrary. Hard bases are ligands not polarizable and electronegative. Soft bases are polarized with medium electronegativity.

The rule is that hard acids prefer to bind to hard bases and soft acids prefer to bind to soft bases.

This rule helps to rationalize and predict the relative stability of transition metal complexes.

The trends for acids are:
Hard acids: Mg^{2+}, Ca^{2+}, VO^{2+}, Al^{3+}, Sc^{3+}, Cr^{3+}, Ti^{4+}
Soft acids: Cu^+, Ag^+, Au^+, Tl^+, Hg_2^{2+}, Pd^{2+}, Pt^{2+}, Hg^{2+}, Cd^{2+}
Intermediate acids: Mn^{2+} Fe^{2+}, Fe^{3+}, Ni^{2+}, Co^{2+}, Cu^{2+}, Zn^{2+}

The trends for bases (ligands) are:
Hard bases: F^-, O, N, Cl^-, OH^-, NH_3, RNH_2, H_2O, CO_3^{2-}
Soft bases: I^-, S, P, CO, CN^-, SCN^-, R_2S, RS^-, $S=C(NH_2)_2$
Borderline bases include pyridine, aniline, SO_4^{2-}

As an illustration of this concept, thiols, which are soft bases, form strong complexes with soft acids such as Hg^{2+}, Ag^+, Pd^{2+}, Pt^{2+}, Cd^{2+}.

Another example is the cyanide complexes (soft base) of Au^+, Cd^{2+}, Cu^+, Ag^+, Pd^{2+}, Pt^{2+}, Hg^{2+} (soft acids) but also Co^{2+}, Fe^{2+}, Fe^{3+}, Ni^{2+}, Zn^{2+} (intermediate acids).

The sulfate ion can act as a ligand where one oxygen (monodentate) or two oxygens (bidentate) are attached to the metal as for example observed with complexes with UO_2^{2+} (Hennig *et al*, 2007).

Ligands that use only one electron pair for bonding to a metal are called unidentates. There are ligands which use two electron pairs to bind with two orbitals of the metal. These ligands are called bidentates. Such an example is ethylenediamine (en):

$$H_2N\text{-}CH_2\text{-}CH_2\text{-}NH_2$$

Another example is the oxalate ion:

$$\begin{array}{cc} O & O \\ \parallel & \parallel \\ {}^-O\text{-}C\text{-}&C\text{-}O^- \end{array}$$

An example of tridentate ligand is the iminodiacetic acid (IDA):

$$NH\begin{cases} CH_2COOH \\ CH_2COOH \end{cases}$$

forming complexes with two oxygen and one nitrogen electron pairs:

27

$$\begin{array}{c} \text{CH}_2\text{COO}^- \\ \diagup \quad | \\ \text{HN} \longrightarrow \text{M} \\ \diagdown \quad | \\ \text{CH}_2\text{COO}^- \end{array}$$

The pKa values of the two carboxylic groups of IDA are $pK_{a1}=2.38$ and $pK_{a2}=9.33$ therefore at pH <2 the IDA molecule would be fully protonated, at pH>10 would be fully deprotonated while at intermediate pH, the N would be protonated along with one of the two carboxylic groups. The formation constants K_f of the complexes of some metals with IDA would depend on the pH since the binding of the metal would be more favorable as pH increases where the lone pairs of N and O would be easier available. Some values of formation constants of IDA are given below:

Metal	K_f
Sr^{2+}	$1.7*10^2$
Ca^{2+}	$3.89*10^2$
Mg^{2+}	$8.71*10^2$
Cd^{2+}	$5.37*10^5$
Fe^{2+}	$6.31*10^5$
Co^{2+}	$8.91*10^6$
Pb^{2+}	$2.82*10^7$
Al^{3+}	$1.45*10^8$
Ni^{2+}	$7.94*10^8$
UO_2^{2+}	$8.51*10^8$
Fe^{3+}	$2.63*10^{10}$
Cu^{2+}	$4.27*10^{10}$
Hg^{2+}	$5*10^{11}$

An example of hexadentate (6 lone pairs) ligand is the ethylene diamine tetraacetate (EDTA^{4-}):

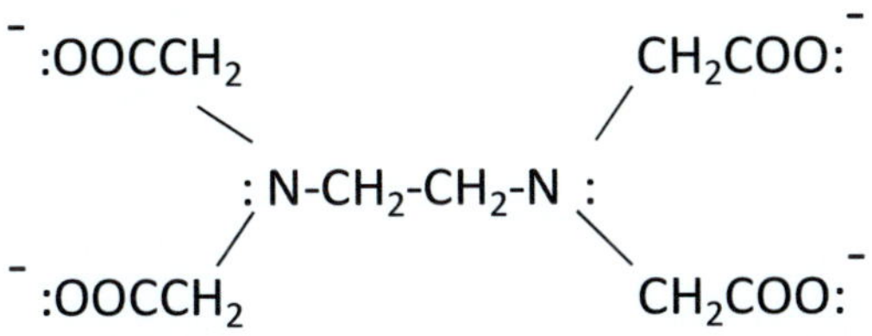

$$\begin{array}{c}
\text{:OOCCH}_2 \diagdown \qquad\qquad \diagup \text{CH}_2\text{COO:} \\
\text{:N-CH}_2\text{-CH}_2\text{-N:} \\
\text{:OOCCH}_2 \diagup \qquad\qquad \diagdown \text{CH}_2\text{COO:}
\end{array}$$

EDTA^{4-} usually forms 1:1 metal-to-EDTA complexes where it acts as hexadentate using two amine and four carboxylate electron pairs, as illustrated in the following figure:

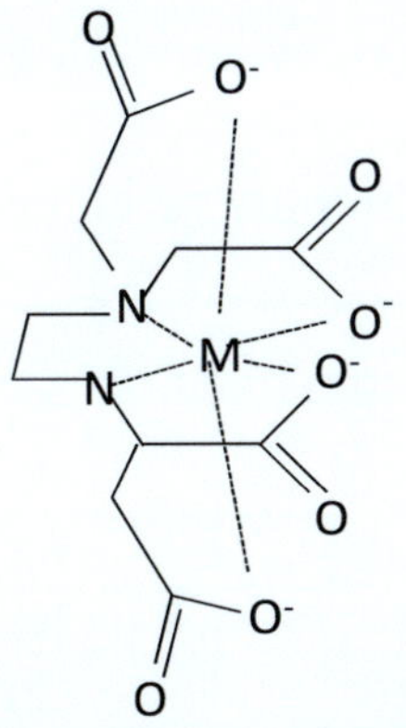

With some metals, EDTA acts as pentadentate (one carboxylic group does not form bond with the metal but it remains with one water molecule).

EDTA is only slowly dissolved in water and has limited solubility. It can be dissolved by neutralizing with NaOH to a pH of 8-8.5, where the solubility is about 0.5 M.

By denoting H_4Y the EDTA molecule in the fully protonated form, the four dissociation constants are:

$$H_4Y \rightleftharpoons H_3Y^- + H^+ \qquad pK_{a1}=2.0$$

$$H_3Y^- \rightleftharpoons H_2Y^{2-} + H^+ \qquad pK_{a2}=2.7$$

$$H_2Y^{2-} \rightleftharpoons HY^{3-} + H^+ \qquad pK_{a3}=6.2$$

$$HY^{3-} \rightleftharpoons Y^{4-} + H^+ \qquad pK_{a4}=10.3$$

The equilibrium, or formation, constants K_f, are in general very large and therefore the following reaction goes very much to the right:

$$M^{n+} + Y^{4-} \rightleftharpoons MY^{n-4} \qquad K_f = (MY^{n-4})/(M^{n+})(Y^{4-})$$

Some examples of K_f values are given below:

Metal	K_f	Metal	K_f
Ba^{2+}	$7.6*10^7$	Zn^{2+}	$3.2*10^{16}$
Sr^{2+}	$4.3*10^8$	Pb^{2+}	$1.1*10^{18}$
Mg^{2+}	$6.2*10^8$	Sn^{2+}	$2.0*10^{18}$
Ca^{2+}	$5.0*10^{10}$	Ni^{2+}	$4.2*10^{18}$
Mn^{2+}	$6.2*10^{13}$	Cu^{2+}	$6.3*10^{18}$
Fe^{2+}	$2.1*10^{14}$	Hg^{2+}	$6.2*10^{21}$
Al^{3+}	$1.3*10^{16}$	Cr^{3+}	$2.5*10^{23}$
Co^{2+}	$2.0*10^{16}$	Sb^{3+}	$6.3*10^{24}$
Cd^{2+}	$2.9*10^{16}$	Fe^{3+}	$1.3*10^{25}$

The complexes with polydentate ligands are called chelating complexes, from the greek χηλή (*chele*) meaning claw and the ligands with two or more electron pairs to donate to the metal ion are called chelating agents. In general, a chelating agent with more than two electron pairs to donate forms stronger complexes than ligands with only one electron pair (*chelate effect*). For example, one molecule of ethylenediamine (en) binds to a metal ion stronger than two molecules of ammonia, the same as two molecules of ethylenediamine bind stronger than four molecules of ammonia.

The stability constants of $[Cu(H_2O)_4(NH_3)_2]^{2+}$ is log K_f =7.86 while for $[Cu(H_2O)_4(en)]^{2+}$ is log K_f =10.86.
This is an entropy driven effect:

$$[M(H_2O)_4(NH_3)_2]^{n+} + en \rightleftarrows [M(H_2O)_4en]^{n+} + 2\,NH_3 \quad (2.1)$$

with $\quad\quad \Delta G° = -RT\ln K = \Delta H° - T\Delta S° \quad\quad (2.2)$

when two molecules of NH_3 are replaced by one molecule of en there is a positive change in entropy. Since the molecules of NH_3 and en are similar, the enthalpy change should be small which makes that the free energy change is negative and therefore the reaction (2.1) above is spontaneous with K>1. Similarly, for the same reason, the formation constants of EDTA are much higher than the corresponding formation constants of IDA (page 28).

Of particular interest for some of the applications that will be discussed later are the complexes of mercury and other metals with sulfur-containing ligands. Thiourea is one of them. It occurs in two tautomeric forms shown below:

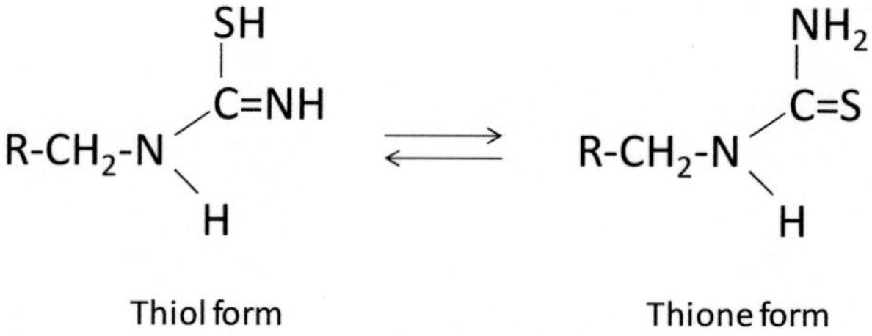

$$R-CH_2-N\begin{matrix}C(SH)=NH\\H\end{matrix} \rightleftharpoons R-CH_2-N\begin{matrix}C(NH_2)=S\\H\end{matrix}$$

Thiol form   Thione form

Thiourea and its derivatives form complexes with a variety of metals such as Ni(II), Pd(II), Pt(II), Hg(II), Zn(II), Sn(II) and Co(II).

Thiols, sometimes called mercaptans (from the latin *mercurium captans,* capturing mercury) are weak acids, stronger than alcohols. For example, thiophenol has a pKa of about 6 while phenol has a pKa about 10. Thus, in the presence of NaOH, thiols form thiolates:

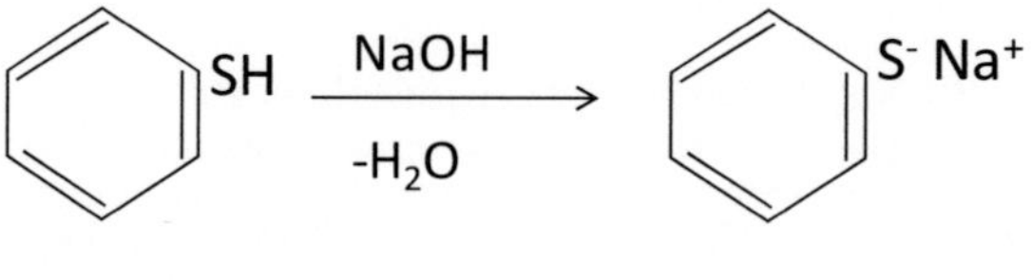

$$Ph-SH \xrightarrow[-H_2O]{NaOH} Ph-S^-\,Na^+$$

The thiols are "soft" acids and form complexes with "soft" bases. Thus, thiolates form strong complexes with "soft" metals such as Hg(II), Ag(I), Cu(II), Pb(II), Cd(II), Ni(II) and Co(II) as shown below for Hg(II):

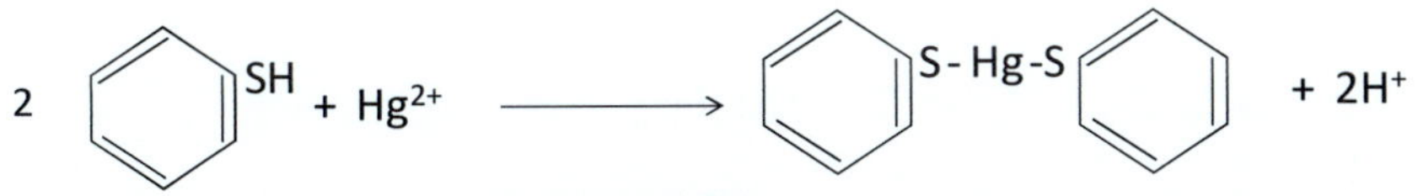

2 [C$_6$H$_5$-SH] + Hg^{2+} ⟶ C$_6$H$_5$-S-Hg-S-C$_6$H$_5$ + 2H$^+$

2:1 complex

The thiol group of cysteine serves as donor atom for coordination bonds with metals such as Hg^{2+}, Pb^{2+} and Cd^{2+} and proteins and peptides containing this aminoacid play important role in the protection of plants or animals from cellular damage from these metals.

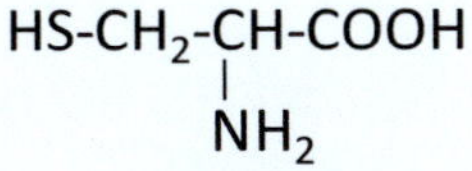

HS-CH$_2$-CH(NH$_2$)-COOH

Cysteine

Among the phosphorous containing ligands where the P atom is the donor atom (soft base) are the phosphines, such as triphenylphosphine, P(Ph)$_3$. The metal phosphine complexes are used often as homogenious catalysts.

Phosphorous containing ligands like phosphonic, HP(=O)(OH)$_2$ and phosphinic, H$_2$P(=O)OH, acids or esters or phosphonates which are organophosphorous compounds containing N atoms, form metal complexes where the O atoms of the phosphonic groups along with the N atom are the donor atoms, in an analogous way as the IDA complexes.

Polyhydric alcohols, like fructose, are also potential ligands for certain metals such as Ca^{2+}, Pb^{2+} or Zn^{2+}.

Boron in aqueous solutions at neutral pH is found as boric acid which is practically undissociated due to its high pKa:

$$B(OH)_3 + H_2O \rightleftarrows B(OH)_4^- + H^+ \quad pKa=9.1$$

Boric acid reacts with polyols to form anionic complexes at neutral pH (Geffen et al, 2006):

$$B(OH)_3 + 2 \begin{array}{c} -C-OH \\ -C-OH \end{array} \rightleftarrows \begin{array}{c} -C-O \\ -C-O \end{array} \!\! B \!\! \begin{array}{c} O-C- \\ O-C- \end{array}^{-} + 3 H_2O + H^+$$

The stability of this complex depends on the type of polyol and, as the above reaction implies, on the pH. At low pH, the reaction goes to the left thus dissociating the boron complex.

Certain oxoanions form complexes with polyols if the hydroxyl groups have favorable steric position (Schilde and Uhlemann, 1993, Parschova et al, 2009) as illustrated above with boric acid. Germanic acid, $Ge(OH)_4$, forms also similar complexes with polyols as boric acid.

Oxoanions, of the formula $A_xO_y^{z-}$ where A represents a majority of elements such as P, B, S, As, Cr, can form complexes with metals (Brintzinger and Hester, 1966). One such example is the complex formation of arseniates with Fe^{3+} (Chanda et al,

1988a). Fe^{3+} was loaded on a chelating resin and the Fe^{3+}-loaded resin was used to remove arseniates from water.

Phenolate anions of various phenolic compounds can act as ligands to form complexes with metal cations such as Fe (Cheng and Crisosto, 1997; Palaniandavar *et al*, 2006), VO^{2+}, UO_2^{2+}, Co^{2+}, Ni^{2+} and Cu^{2+} (Ahmad *et al*, 1994).

It is interesting to note that metal ions can interact with polymeric adsorbents to form complexes. Cation-π interaction (Mecozzi *et al*, 1996), a non-covalent binding force that binds a cation to the π surface of an aromatic ring, is probably the mechanism for complex formation between a metal and the benzene ring of styrene/DVB adsorbents (Davankov *et al*, 2002). It has also been found that certain metals in acidic media can be adsorbed on acrylic adsorbents such as Amberlite® XAD7 (Koshima, 1986; Harris *et al*, 1991; Harris and White, 2012). The mechanism is possibly the interaction of the chlorocomplexes with the oxygen atoms of the acrylic adsorbent.

The formation of complexes of metal cations with various ligands affects the mechanism of ion exchange in a number of ways:

- Complex formation in the solution phase between a metal cation and a ligand affects the distribution of the metal cationic species between the ion exchange resin and the solution. For example in the removal of Zn^{+2} from HCl solutions with an anion exchange resin, Zn^{2+} forms the $[ZnCl_4]^{2-}$ complex which is not excluded by the Donnan effect and can then enter the resin where it is fixed by exchange with the counter-ion (Helfferich 1962).

- Complex formation between a ligand in the solution and a metal cation found as counter-ion on a cation exchange resin, for example removal of amines or of ammonia from aqueous solutions with a weak acid cation exchanger in the Cu^{2+} form (Bolden and Groves Jr, 1990). Ligand exchange is based on this mechanism (Helfferich, 1962). In the above example, ammonia is exchanged against water molecules from the solvation shell of Cu^{2+}. The ligand exchanger (that is, the cation exchange resin in the form of a complexing cation) can act as a selective and high capacity sorbent for the ligand. In the above example, Cu^{2+} has a coordination number of four and can therefore fix four molecules of ammonia per Cu^{2+} atom:

$$(R^-)_2 [Cu(H_2O)_4^{2+}] + 4\,NH_3 \leftrightarrows (R^-)_2 [Cu(NH_3)_4^{2+}] + 4\,H_2O$$

- Complex formation between a ligand found on an ion exchanger as counter-ion and a metal cation in solution, for example removal of Ag^+ from solutions with an anion exchange resin in the SCN^- form (Kononova *et al*, 2006).
- Complex formation between a ligand bound on an ion exchange resin with a covalent bond as a functional group and a metal cation found in the solution. Ion exchangers bearing complexing agents (ligands) as functional groups are widely used in removing metal cations or other complex-forming ions from aqueous solutions (Trochimczuk and Jezierska, 2000). Such resins have a polymer matrix bearing, among other, the following functional groups: polyamine (tri- or tetraethyleneimines), iminodiacetic,

aminomethylphosphonic, di-(2-picolylamine), glucamine or thiol. Ligand exchange in this case can also take place as described above, for example removal of arsenates from solutions with a chelating resin in the Fe^{3+} form (Chanda *et al*, 1988a).

Regeneration of the resins in the above cases can be done either by adjusting the pH so that the complex becomes unstable, or eluting the metal using another ligand which forms a stronger complex with the metal or by simple ion exchange.

Ion exchange resins with ligand functional groups

A number of ion exchange resins have been synthesized where a ligands such as those described above, are bound on the polymer matrix of the resin by a covalent bond. These resins are known since the years 1950's (Pennington and Williams, 1959). Resins with bi- or polydentate ligands are also called chelating resins. Following are some of these resins frequently used in the chemical industry:

Weak base anion exchange resins

It is sometimes overlooked that conventional WBA exchange resins having primary, secondary or tertiary amine groups, have the ability at about neutral or alkaline pH to donate one electron pair of the N atom to a metal to form a coordination complex. This allows for example these resins to remove transition metals from water (Höll, 1996) without an ion exchange mechanism.

Regeneration in this case is done with acids where the proton of the acid displaces the metal from the N atom of the functional group.

Iminodiacetic acid resins

The structure of these resins is shown below:

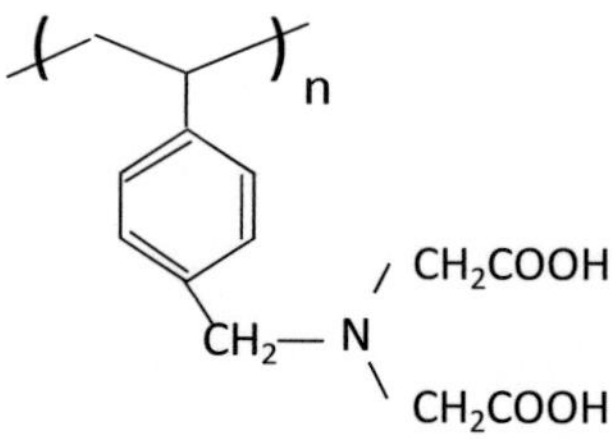

$$\left(\begin{array}{c}\\\end{array}\right)_n - C_6H_4 - CH_2 - N\begin{array}{c}- CH_2COOH\\- CH_2COOH\end{array}$$

The functional group of this resin is the iminodiacetic acid (IDA) which is the analog of the IDA ligand where the nitrogen proton has been replaced by a carbon atom. This resin forms complexes in a similar way as IDA (page 28). Three coordination numbers of the metal share the electron pairs of the two donor oxygen atoms and the nitrogen atom, and also the positive charges of the metal are neutralized by the negative charges of the acetate groups, thus both electrostatic interaction and Lewis acid-base interaction bind the metal ions on the IDA type resins:

[Structure: polymer backbone with phenyl group bearing CH_2-N with two $CH_2COO^:-$ groups coordinating to M^{2+}]

By forming complexes with metal cations in solution, this type of resins shows a very high selectivity for complex forming metals with respect to elements that do not. For example, these resins can remove traces of metal cations such as Ni^{2+}, Fe^{3+}, Cu^{2+} even Ca^{2+} or Sr^{2+}, from solutions of 30% NaCl where a conventional strong acid cation resin would not work.

The pKa values of the carboxylic groups of IDA resins are about $pKa_1=3.3$ and $pKa_2=8.4$ while for the amine is $pKa_3=9.6-10.4$. Figure 2.1 shows the zwitterionic forms of the resin at various pH. At very low pH (fig. 2.1 A), the carboxylic and amine groups are fully protonated and the IDA type resins behave in fact as anion exchangers:

(A) [polymer with CH_2-NH^+ bearing two CH_2COOH groups]   (B) [polymer with CH_2-NH^+ bearing CH_2COO^- and CH_2COOH]

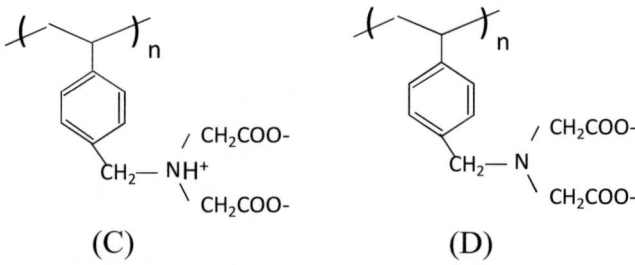

Figure 2.1 Iminodiacetic acid resins

At a pH between about 4 and 8, one or both of the carboxylic groups are deprotonated (fig. 2.1 B and C). At a pH above about 10, the groups are fully deprotonated (fig 2.1 D):

In order that the metal ion forms complexes with IDA resins, the resin should be found in the deprotonated form (figure 2.1D). However, metals have the ability to displace the protons from the carboxylic or the amine groups and they can form complexes at pH values more or less acidic. The different metals have different abilities to displace protons from the functional groups and therefore, the pH can differentiate the metals in forming complexes with IDA resins.

The approximate selectivity sequence of these resins is as follows (Neumann, 2008):

$Fe^{3+} > Cu^{2+} > H^+ > Hg^{2+} > Pb^{2+} > Ni^{2+} > Zn^{2+} > Cd^{2+} > Co^{2+} > Fe^{2+} > Mn^{2+} > > Ca^{2+} > Mg^{2+} > Sr^{2+} > Ba^{2+} >>>> Na^+$

These selectivities are qualitatively in the same order as the formation constants of the pure chelating agent IDA mentioned in page 28. In a solution containing Ca^{2+}, Mg^{2+} and metals such as Pb^{2+}, Ni^{2+}, Zn^{2+} etc, the resin still prefers the metals with respect to Ca^{2+} or Mg^{2+}, however, the equilibrium constant becomes lower than in the absence of the hardness elements. This should results to a lower operating capacity of the resin to remove metals from high hardness solutions compared to solutions containing only alkali metals.

IDA resins fix strongly Cr^{3+}, the regeneration however is difficult. Hot (50°C) regeneration with HCl allows the practically complete elution of Cr^{3+}.

In some cases the above selectivity sequence is not strictly comparable with the formation constants sequence of page 28 and this is explained by a different complexation of the IDA groups of the resin (Pesavento *et al*, 1993) or by a "polymer effect" where the polymer structure affects the formation of the metal complexes. For example, if we write HL the monoprotonated iminodiacetate, at pH=5 Ca^{2+} is complexed with two iminodiacetate groups:

$$Ca + 2HL \leftrightarrows Ca(HL)_2$$

while at pH=8 it form complexes with one iminodiacetate group:

$$Ca + HL \leftrightarrows ML + H$$

Cu^{2+} and Ni^{2+} form complexes with one IDA group at pH=5. With Cd^{2+} and Zn^{2+} the above two possibilities coexist.

It should be mentioned that, depending on the manufacturing process, the commercial IDA type resins may have some of their functional groups in the mono-acetate form. These resins would maintain their high selectivity for metals with respect to alkali metals, however it may affect the values of the formation constants of the complexes compared to the IDA ligand. On the other hand, the kinetics of these resins may be faster so that they give a sharper breakthrough curve.

It is pointed out in the above selectivity sequence that H^+ is found high and between Cu^{2+} and Hg^{2+}. This implies that at low pH values, the resin will be in the protonated form and will not be able to fix significantly the metal cations found to the right of H^+ in the above sequence. On the contrary, even at pH values where the resin is protonated, Fe^{3+} and Cu^{2+} can be fixed by the resin. The resin before put into service to fix metals on the right hand of H^+ in the above sequence, is converted partially to the sodium form, the rest being in the H^+ form. For the elements Fe^{3+} and Cu^{2+}, the resin can be in the H^+ form.

The reason for converting the resin partially to the Na^+ form is that due to the hydrolysis of the functional groups in Na^+ form, the effluent pH can be alkaline thus provoking a metal precipitation. The effluent pH is a function of the degree of conversion of the resin to the Na+ form and on the ionic background of the feed solution. With the resin fully in the Na^+ form, rinsing with water gives an initial effluent pH of around 12. With high ionic background Na^+ and Ca^{2+} salts it is slightly lower. At 50% conversion to the Na^+ form, the initial pH is about 8, lower for a background with Na^+ and Ca^{2+} salts.

The selectivity as well as the operating capacity of these resins in fixing metals depends on the pH of the solution.
The operating capacity vs. pH curve for the above metals is an S-shaped curve: at low pH the resin operating capacity is very low, at a certain pH the curve increases rapidly to reach a high

value which remains constant at further increase of the pH (until a pH is reached where the metal precipitates out). The pH at which the operating capacity starts going up is not the same for all the above metals but increases as the metal selectivity decreases. This is illustrated in the following table:

Table

pH	1.5	2	2.5	3	3.5	4	5
Metal	Fe^{3+}	Cu^{2+}	Pb^{2+}	Ni^{2+}	Cd^{2+} Zn^{2+} Co^{2+}	Al^{3+}	Mn^{2+} Ca^{2+}

Another factor that affects the selectivity sequence is the presence of other ligands in the solution. Thus, Hg^{2+} is found high in the selectivity sequence in the presence of NO_3^- ions but low in the presence of Cl^- ions with which it forms complexes. For the same reason, IDA type resins cannot remove metals from solutions in the presence of the EDTA complexing agent since the formation constants of EDTA is much higher than those of IDA (pages 28 and 30). IDA resins cannot remove Cu^{2+}, Ni^{2+} or Zn^{2+} found in solution as cyanide complexes.

The effect of NH_3 on the selectivity of IDA resins (Amberlite™ IRC748) is illustrated by the following numbers, at a pH of 9 in a high concentration of $(NH_4)_2SO_4$ background (Rohm and Haas, 2001). In particular, Cu^{2+} that is high in the selectivity sequence at low pH, in presence of high NH_4^+ concentration is lower than Ni^{2+} or Cd^{2+} due to its strong complexes with ammonia:

Metal	K=M/Ca
Ni^{2+}	30
Cd^{2+}	14
Cu^{2+}	10
Zn^{2+}	3
Ca^{2+}	1

Commercial resins of IDA type include Amberlite™ IRC748, Lewatit™ TP207 (for metal recovery) and Lewatit™ TP208 (for brine purification), Purolite™ S930 and Diaion™ CR-11.

Aminomethylphosphonic resins

These resins have the following structure:

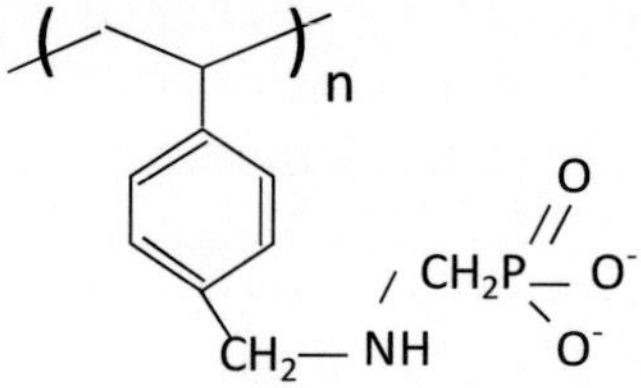

$$\left(\begin{array}{c} \\ \end{array} \right)_n \underset{CH_2-NH}{\overset{}{\bigcirc}} \diagdown CH_2-\underset{O^-}{\overset{O}{\underset{\|}{P}}}-O^-$$

The functional group corresponds to the ligand aminomethylphosphonic (AMP) acid:

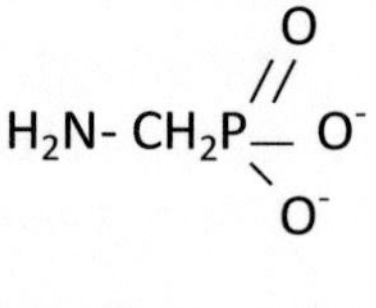

$$H_2N-CH_2-\underset{O^-}{\overset{O}{\underset{\|}{P}}}-O^-$$

which has been found to form stable complexes with alkaline earth and polyvalent metal cations.

The resin forms tridentate complexes with two oxygen and one nitrogen electron pairs:

[Chemical structure: polymer backbone with phenyl group bearing $CH_2-NH-CH_2P(=O)(O^-)(O^-)$ coordinated to M]

The selectivity sequence of AMP resins for different metals is:

$UO_2^{2+} > Pb^{2+} > Cu^{2+} > Zn^{2+} > Ni^{2+} > Cd^{2+} > Co^{2+} > Ca^{2+} > Mg^{2+} > Sr^{2+} > Ba^{2+} > Na^+$

Compared to IDA type resins, the AMP type shows higher selectivities for elements of low atomic mass, for example Mg^{2+} and Ca^{2+}. The reverse is true for the elements Sr^{2+} and Ba^{2+}: IDA type resins have higher selectivities for these elements compared to AMP type resins. This is taken into account in choosing a resin for purifying brine in the chloralkali industry. Depending on whether the criterion for purity of the brine is Sr^{2+} and Ba^{2+} or Ca^{2+} and Mg^{2+}, the resin choice would be IDA type or AMP type respectively.

Commercial resins of AMP type include Amberlite™ IRC747, Duolite™ C467, Lewatit™ TP260 and Purolite™ S940 and Purolite™ S950.

Other phosphorus containing resins

Resins having phosphoric, phosphonic, phosphinic and thiophosphinic acid functional groups similar to organophosphorus liquid extractants (such as Cyanex® liquid extractants), have been synthesized:

Phosphoric acid — P(=O)(OH)(OH)(OR)

Phosphonic acid — P(=O)(OH)(OH)(R)

Phosphinic acid — P(=O)(H)(OH)(R)

Thiophosphinic acid — P(=S)(H)(HS)(R)

R = polymer backbone

The complexes with metals are formed with the O (or S) atoms of the deprotonated OH groups and with the =O atom. Resins having a styrene/DVB matrix with diphosphonic, monophosphonic and phosphinic functional groups where sulfonic groups have been added have been synthesized. In fact, the sulfonic groups improve the accessibility of the phosphinic groups which in turn are responsible for the complex formation and therefore for the selective removal of metals:

Diphonix® resins (from Eichrom Industries) belong to this type of resins (Alexandratos *et al*, 1994; Chiariza *et al*, 1997; Horwitz *et al*, 2002). The diphosphonic groups provide high selectivity for selected metals while the sulfonic groups provide hydrophilicity thus enhancing the accessibility of the diphosphonic groups and improving the kinetics of the resin. A similar effect was observed with bifunctional SBA exchange resins, selective for perchlorates and pertecnetates (Alexandratos *et al*, 2000), mentioned earlier (page 22). The Diphonix® resins consist of a styrene/DVB matrix on which are grafted the phosphonic and the sulfonic groups. There exist also the Diphosil® resins where the diphosphonic groups are grafted on a silica support. Diphosil® resins have therefore a much lower level of carbon-hydrogen bonds than Diphonix® and are better suited for treating radioactive wastes.

Diphonix® resins are selective for metal cations such as iron, zinc, manganese, chromium, uranium, lead, nickel and cobalt. They also exhibit high selectivity for lanthanide and actinide ions in various oxidation states from high acidic solutions. Monophosphonic resins show high selectivity for Fe^{3+} and is used in removing Fe^{3+} from Cu^{2+}, Ni^{2+} or Co^{2+} electrolytes.

Purolite® S-957 (from Purolite International Ltd), Tulsion® CH-96 (from Thermax Limited) are among the commercial resins of this type.

Another commercially available resin is Diaion® CRP200, bearing methylenephosphonic groups:

$$\left[\begin{array}{c}\text{—CH}_2\text{—CH—}\\|\\\text{C}_6\text{H}_4\\|\\\text{CH}_2\text{P(=O)(O}^-\text{)}_2\end{array}\right]_n$$

There exist also solvent impregnated resins (SIR) with di-2-ethylhexylphosphoric acid or bis-(2,4,4-trimethylpentyl-)phosphinic acid (p. 63) to remove cobalt from nickel electrolytes.

Picolylamine resins

di-(2-picolylamine) (or bis-(2-pyridylmethyl)amine) is a tridentate ligand that can coordinate with transition metal cations through one amine and two pyridine nitrogens:

Bis-picolylamine (BPA) resins have a polystyrene-DVB copolymer matrix where a bis-picolylamine molecule is bound by replacing the proton of the amine nitrogen by a carbon atom:

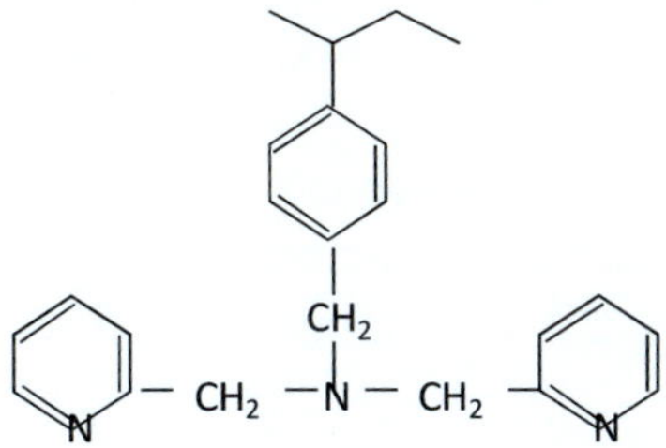

BPA

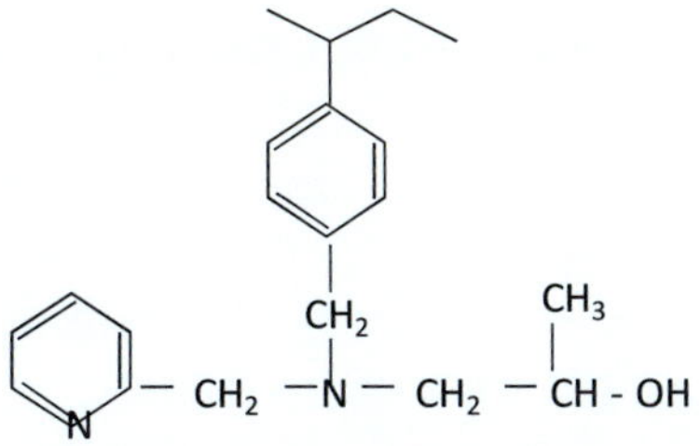

HPPA

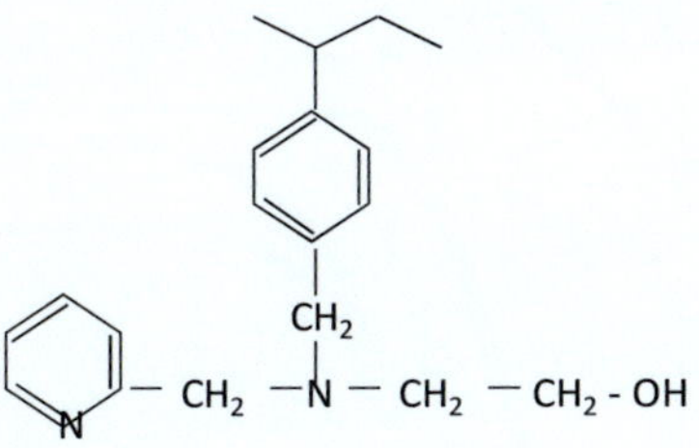

HEPA

Another resin of the same type has N-2-hydroxypropyl)-picolylamine (HPPA) functional groups (Grinstead, 1979). This resin showed a higher Cu/Fe(III) selectivity than BPA. BPA and HPPA resins have strong affinity even at low pH (<2) for transition metals in the following order (at pH=2):

$$Cu^{2+} \gg Ni^{2+} > Fe^{3+} > Zn^{2+} > Co^{2+} > Cd^{2+} > Fe^{2+}$$

Another resin developed but not any longer available commercially was the Dowex XFS 4196, with N-(2-hydroxyethyl)-picolylamine (HEPA) functionality (Grinstead, 1979). The characteristic of this resin was that it could be regenerated with dilute sulfuric acid.

These resins can capture transition metals in presence of strong complexing agents such as EDTA. Figure 2.2 illustrates the relative selectivities of Dowex® M-4195, a BPA resin, as a function of pH for various metals.

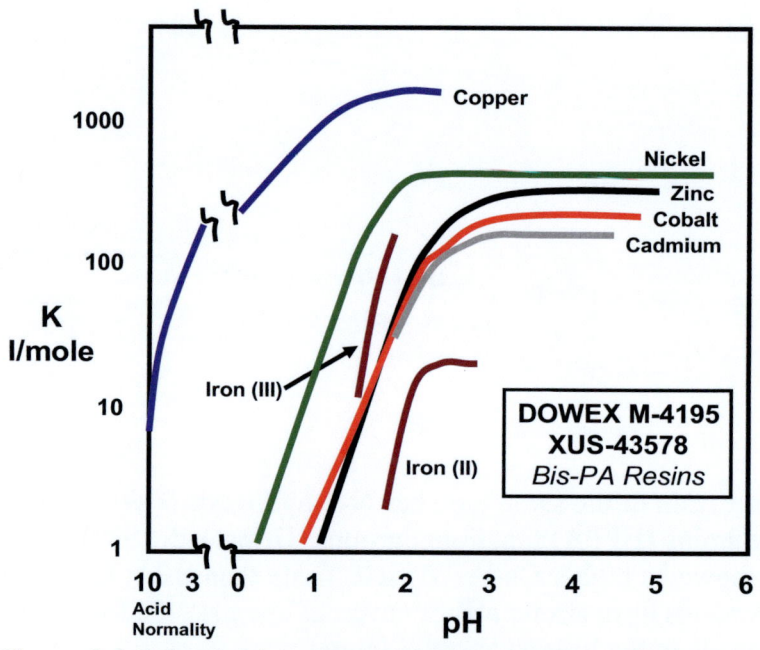

Figure 2.2 Selectivities of Dowex® M-4195 as a function of pH (Courtesy of The Dow Chemical Company)

One commercial application that these resins have found is the cobalt electrolyte purification where Dowex® M-4195, a BPA resin, was used to remove nickel at very low pH (Kotze, 2012). Another application of BPA resins is the purification of Cr(III) plating baths by removing copper and nickel impurities.

HPPA resins introduced by The Dow Chemical Company differ in that they load less strongly copper allowing an elution with less concentrated H_2SO_4. In addition, they have more iron rejection properties. Figure 2.3 illustrates the relative selectivities for Dowex® XUS 43605 (XFS 43084) as a function of pH.

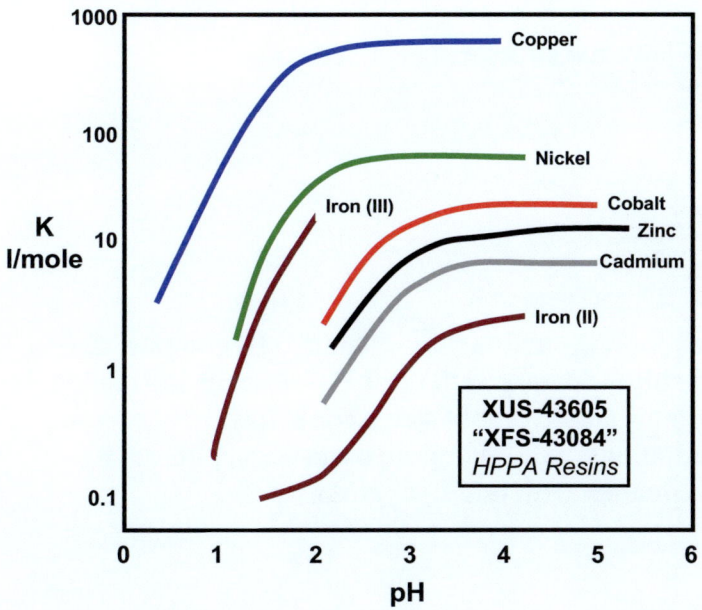

Figure 2.3 Selectivities of Dowex® XUS 43605 as a function of pH (Courtesy of The Dow Chemical Company)

Commercial products of BPA resins include Dowex® M-4195, Purolite® S960 and Lewatit® TP220. Commercial product of HPPA resins is Dowex® XUS 43605.

Regeneration of these resins is done with H_2SO_4 or NH_4OH. In fact, it is possible by applying a split elution with H_2SO_4 at different strengths to separate metals fixed by the resin. Or else, with an elution with 5% H_2SO_4 to elute metals such as Zn^{2+}, Co^{2+}, Ni^{2+}, Fe^{3+}, and then with NH_4OH to elute copper.

Thiol resins

Thiol resins have the structures depicted below:

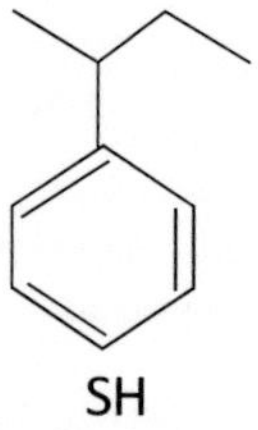

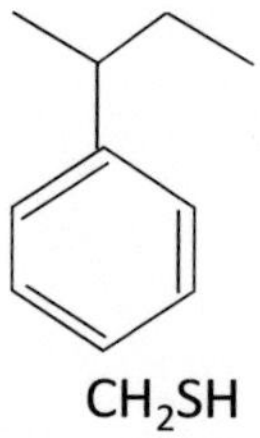

SH CH_2SH

Thiol resins form the stable thiol-mercury (II) complex due to the favored soft S-donor and the soft Hg-acceptor interaction. If two thiol groups are favorably spaced or at low Hg^{2+} concentrations, two thiol groups can form a linear two-coordinate complex with one Hg^{2+} atom.

S- Hg²⁺Cl⁻   S- Hg -S

1:1 complex   2:1 complex

These are resins developed to remove traces of mercury from waste solutions. At the same time, they have high selectivity for other metals like Ag^+, Cu^{2+}, Pb^{2+}, Cd^{2+} and precious metals. It also reacts with As(III) to form As-S bonds.

Regeneration of the resin is done with concentrated hydrochloric acid. There exist resins which have the thiol group grafted on the benzene ring or via a methylene bridge.

Ambersep® GT74 (from Rohm and Haas Company) contains thiol groups directly grafted on the aromatic ring. It shows the following selectivity sequence (Rohm and Haas Company, 2006):

$Hg^{2+} > Ag^+ > Cu^{2+} > Pb^{2+} > Cd^{2+} > Ni^{2+} > Co^{2+} > Fe^{3+} > Ca^{2+} > Na^+$

Another commercial product having thiol functional group is Purolite® S924. Ionac® SR 4 of Sybron Chemicals Inc is a resin containing methylthiol functional group.

Oxidative agents should be avoided because the resins loose –SH functional groups, and therefore capacity for removing Hg and other metals, by the formation of –S-S- bonds or even $-SO_3^-$ groups. The disulfide is the usual end product of air oxidation:

2 R–SH + ½ O_2 ⟶ R-S-S-R + H_2O

The $-SO_3^-$ groups are formed by stronger oxidizing agents.

Thiourea resins

Thiourea resins structure is shown below:

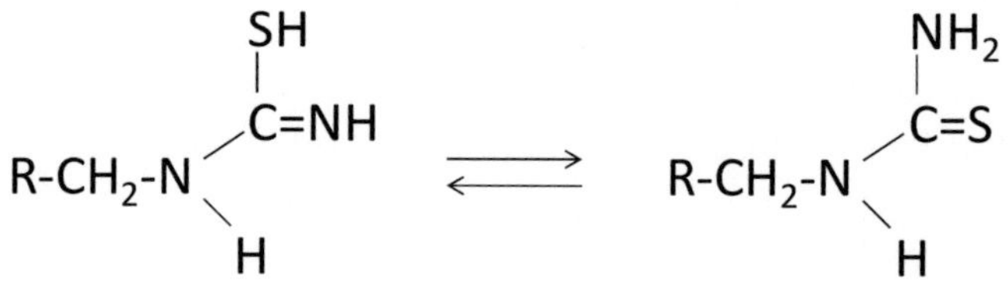

$$R-CH_2-N\begin{matrix}C(SH)=NH\\H\end{matrix} \rightleftharpoons R-CH_2-N\begin{matrix}C(NH_2)=S\\H\end{matrix}$$

Thiol form   Thione form

Thiourea resins

The two N atoms along with the S of the thiones form coordination complexes with metals such as Zn, Cd and Hg. Lewatit® TP 214 of Lanxess, a thiourea type resin, shows the following selectivity sequence (Lanxess, 2011b):

$Hg^{2+} > Ag^+ > Au^{1+/3+} > Pt^{2+/4+} > Cu^{2+} > Pb^{2+/4+} > Bi^{2+} > Sn^{2+} \; Zn^{2+} > Cd^{2+} > Ni^{2+}$

Another commercial product of this type is Purolite® S914. Regeneration of Hg and precious metals is not possible with usual regenerants.

Thiouronium resins

The chemical structure of thiouronium (isothiourea) resins is shown below:

$$R-CH_2-S-C\begin{smallmatrix}NH_2\\NH\end{smallmatrix}$$

Thiouronium resins

It forms complexes with metals such as mercury, gold, silver, platinum and palladium:

$$R-CH_2-S-C\begin{smallmatrix}NH_2\\NH\end{smallmatrix} + Hg^{2+} \longrightarrow R-CH_2-S-C\begin{smallmatrix}NH_2\\NH\end{smallmatrix} \;\; Hg^{2+}$$

Thiouronium resins are more resistant to air oxidation than thiol type resins. Dowex® XUS 43600.00 of The Dow Chemical Company and
Purolite® S-920 of Purolite International Ltd are thiouronium type resins. These resins are unstable (hydrolysis) in alkaline pH and the resin looses the thiouronium functional group (replaced by methylthiol groups) and changes chemical character. Although the alkaline hydrolysis of the thiouronium salt gives methylthiol and urea:

$[R-CH_2-S-C(NH_2)_2]^+X^- + NaOH => R-CH_2-SH + OC(NH_2) + NaX$

if this reaction takes place during the loading cycle, then some metals may leak out by forming complexes with the urea molecule via the oxygen electron donor atom. In addition, it is not clear if the above mechanism of alkaline hydrolysis of thiouronium remains the same when the S and N atoms are bound to metals such as Hg^{2+} by coordination bonds.

Dithiocarbamate resins

Resins with dithiocarbamate functionality form strong complexes with transition metals where the S atoms are the electron donors.

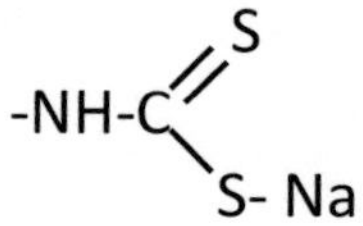

$$-NH-C\underset{S-Na}{\overset{S}{\lessgtr}}$$

These resins show high affinity for Hg^{2+}, Cd^{2+}, Pb^{2+} and precious metals. The affinity of these resins depends on the amine type used to graft the dithuocarbamate group.
Commercial products include Nisso ALM 525 of Nippon Soda Co. The selectivities of this type of product for transition metals are:

$$Hg^{2+} > Cd^{2+} > Zn^{2+} > Pb^{2+} > Cu^{2+} > Ag^{+} > Cr^{3+} > Ni^{2+}$$

Polyamine resins

The polyamine functional groups give these resins the property of forming complexes with transition metals that form stable ammine complexes, such as Ni^{2+}, Co^{2+} or Cu^{2+}. With this mechanism these weak base resins can selectively remove transition metals from aqueous solutions without changing the ionic composition of the solution being treated, that is without ion exchange. Only the metal salt is absotbed by the resin.

Such a polyamine resin can be made from a phenol-formaldehyde resin after reacting with a polyamine like triethylenetetramine and eventually crosslinked with epichlorohydrine:

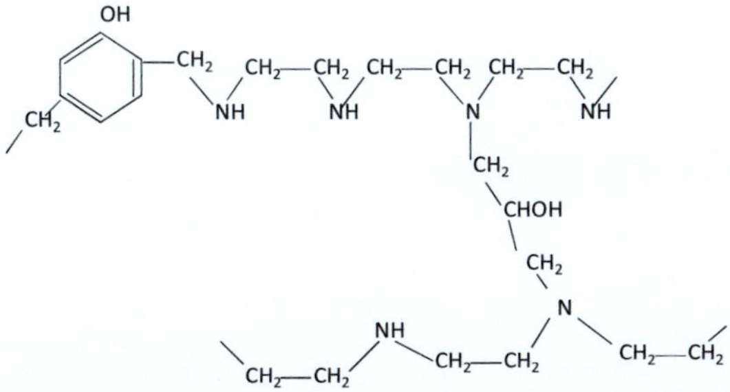

There exist also polyamine resins having an acrylic matrix (Purolite® S-985).

Cross-linked polyethylene imine polymers have also been used immobilized or impregnated on various solid supports:

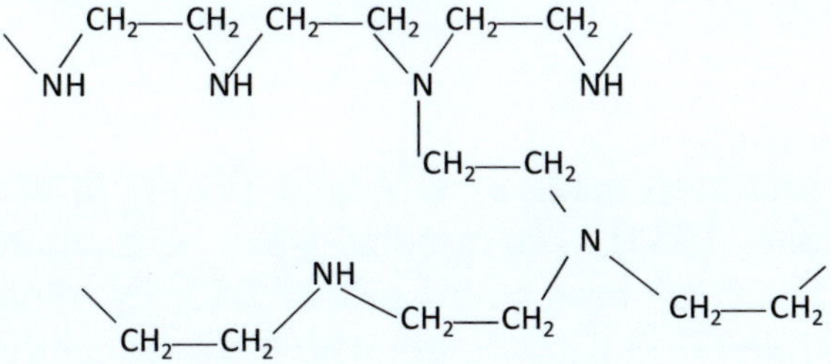

Depending on the pH, the N atoms can be protonated or not, according to the pKa value of the amine. In the protonated form, the formation of metal complexes is inhibited and the resin can act as anion exchange resin. At intermediate pH values, the resin can exist in both protonated form and free base form and can complex with metals and also exchange anions.

Amidoxime resins

Amidoxime resins can be made from polyacrylonitrile/DVB copolymer by reacting with hydroxylamine (Hubicki and Kołodyńska, 2012). The complex of amidoxime groups with metal cations is shown below:

$$P-\underset{NH_2}{C=N-O-H} \qquad P-C\underset{HN-O}{\overset{NH}{\diagup}}M$$

Amidoxime resins have been used to recover uranium from sea water and gallium from the bayer liquors. It shows also affinity for As^{3+}, Cu^{2+}, Pb^{2+}, Cd^{2+} and Fe^{3+}.
Commercial products include Purolite® S910

Pyridine resins

Pyridine resins have first been made by D'Alelio already in the years end-40's (D'Alelio, 1954) to be used as weak base anion exchangers. Pyridine is a weak base with a pK_a of 5.17. Pyridine resins are 4-vinylpyridine/DVB copolymers which subsequently can be quaternized to form a strong base anion exchanger:

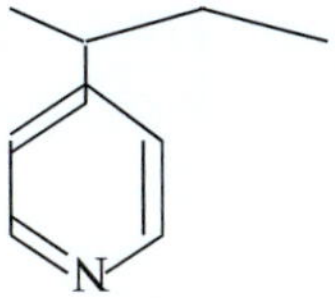

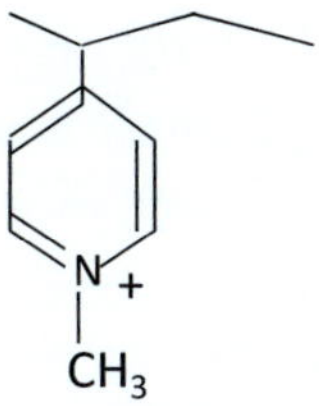

According to the HSAB theory, pyridine is a borderline base and is known to form complexes with metal ions like the PGM, Cu, Ni, Cd. Pyridine resins have shown a better stability to oxidation and improved loadings with PGM from HCl solutions and with uranium from acid leach compared to styrene/DVB SBA resins.

Methylglucamine resins

Ion exchange resins selective for boron are based on the reaction of boric acid with polyols to form complexes (page 34). Resins with N-methylglucamine functional groups are made from styrene-DVB copolymers in a similar way as in the synthesis of conventional weak base resins but using glucamine in the raw materials.

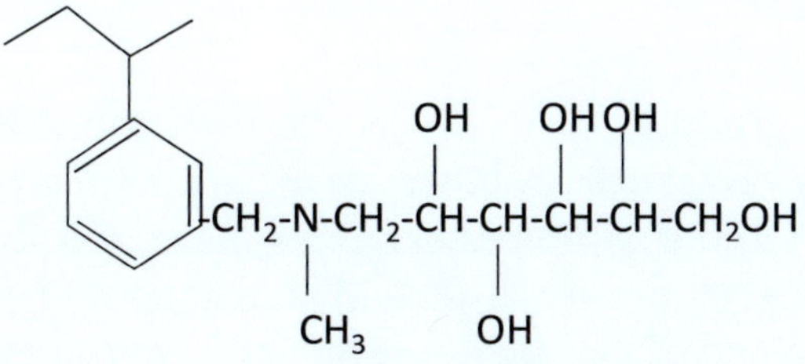

Methyl glucamine resin

The structure of boron selective resins is therefore a WBA bearing a deoxy-sorbitol chain attached to the amino group.

This resin can selectively remove boric acid at pH from slightly acidic to alkaline, from low salinity waters to very concentrated brines (see Chapter 5, p. 202). Being a WBA resin having eventually, depending on the manufacturing process, some strong base groups, it can function, in addition to the boron removal, as an anion exchange resin.

Germanic acid, $Ge(OH)_4$, can be removed from solutions and recovered with methylglucamine resins (Virolainen, 2013) in a similar way as boric acid. Ge was removed from sulfate solutions of Co and Zn using methylglucamine resins (Amberlite IRA743 of Rohm and Haas Company, now The Dow Chemical Company) with capacities varying depending on the pH of the solutions. The recovery of germanium was achieved on the same principles as in the boron removal described in Chapter 5 (page 202).

Although styrene-DVB type resins are typical commercial resins, there exist other products such as resins based on branched polyethylene imine where are grafted boron-selective groups by reacting with glucono-1,5-D-lactone (Mishra et al, 2012).

Carboxylic resins

The carboxylic group of WAC resins can form monodentate complexes with metal atoms, however these complexes are too weak compared to the complexing agents discussed above. Combining electrostatic attraction and coordination bonding, the selectivity sequence of carboxylic resins at a neutral pH is:

$$Pb^{2+} > Al^{3+} > Cr^{3+} > Cu^{2+} > Zn^{2+} = Ni^{2+}$$

Special types of resins

Sodium borohydride, $NaBH_4$, is a strong reducing agent in many applications. Another reducing agent is the amine-boranes, R_3NBH_3, which are milder reducing agents and are more stable and more selective than borohydride. These amine-boranes find use in electroless plating.

Some years ago, Rohm and Haas Company developed polymeric amine-boranes under the trade name of Amboranes®, which are more stable and more selective reducing agents than amine-boranes (Manziek 1982a). Although these products are not complexing resins, they are mentioned here because they were used to recover metals. Their main application was the selective reduction of precious metals such as Au^{3+}, $Pt^{2+/4+}$, Pd^{2+}, Ag^{+}, Ir^{3+}, Rh^{3+}, in the presence of most base metals. The reaction of these resins is represented by:

$$n\ RN\text{-}BH_3 + 6\ M^{n+} + 3n\ H_2O \longrightarrow n\ RNH^{+} + 6\ M^{\circ} + 5n\ H^{+} + n\ B(OH)_3$$

The metal is precipitated within the resin matrix from where it is recovered by roasting the resin. Prior to roasting, any unreacted amine borane groups should be removed by reacting with an aldehyde solution (Manziek, 1982b).

Today, these products are no longer commercially available in the market.

Another class of special resins developed by IBC Advance Technologies and promoted under the name of SuperLig^{TM} is used in *Molecular Recognition Technology (MRT)* applications. These are very selective materials bearing specially designed ligands, such as macrocycles, bonded or not on solid supports such as silica gels or polymers.

Finally, another example of chelating products is liquid extractants adsorbed on polymeric adsorbents which are then used as solid ion exchangers. These products are known as *Solvent Impregnated Resins (SIR)* and contain such solvents as di(2-ethylhexyl) phosphoric acid (D2EHPA) among others.

$$2 \left[\begin{array}{c} CH_3\text{-}(CH_2)_3\text{-}\underset{|}{CH}\text{-}CH_2\text{-}O \\ C_2H_5 \\ CH_3\text{-}(CH_2)_3\text{-}\underset{|}{CH}\text{-}CH_2\text{-}O \\ C_2H_5 \end{array} \right. \underset{P}{\diagdown} \underset{O\text{-}H^+}{\overset{O}{\diagup}} \left. \right] \xrightarrow{Zn^{2+}}$$

D2EHPA

$$\underset{\underset{C_2H_5}{|}}{CH_3\text{-}(CH_2)_3\text{-}CH\text{-}CH_2\text{-}O} \diagdown \underset{P}{\diagup} \overset{O}{\diagdown} \underset{Zn}{\diagup} \overset{O}{\diagdown} \underset{P}{\diagup} \overset{O\text{-}CH_2\text{-}CH\text{-}(CH_2)_3\text{-}CH_3}{\underset{\underset{C_2H_5}{|}}{\diagdown}} + 2H^+$$

SIR containing various solvents have been used to recover metals from aqueous solutions and eventually separate them by selective elution using different eluents (Taute *et al*, 2013). Frequently, SIR are made by the end-users starting from commercially available polymeric adsorbents. Amberlite® XAD4 impregnated with D2EHPA has been reported. Some commercially available SIR are made by incorporating the liquid extractants during the polymerization of the adsorbent rather than impregnating afterwards. Such a product is Lewatit® VP OC 1026 containing D2EHPA (Lanxess, 2012). This technique allows the incorporation of more extractant with fewer losses during operation. The selectivity sequence of D2EHPA impregnated resins is the following:

$Sc^{3+} > Fe^{3+} > UO_2^{2+} > REE > Zn^{2+} > Cd^{2+} > Pb^{2+} > Ca^{2+} > Mn^{2+} > Cu^{2+} > Fe^{2+} > Sr^{2+}, Mg^{2+}, Ba^{2+} > Co^{2+} > Ni^{2+}$

Another similar product is a crosslinked styrene-DVB macroporous resin containing bis-(2,4,4-trimethylpentyl-) phosphinic acid:

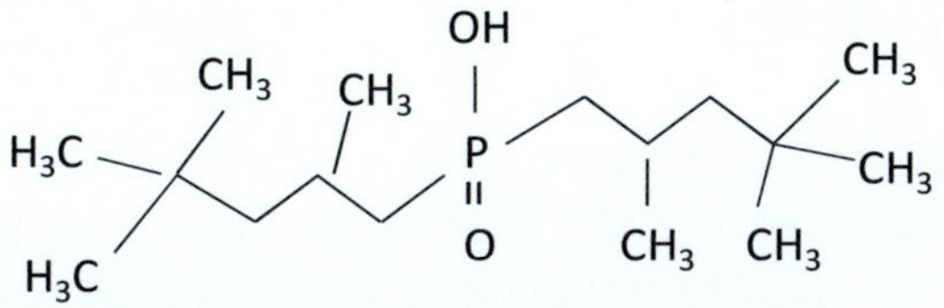

This active ingredient is directly incorporated in the resin by absorption, during the polymerization. The main application is Co^{2+} removal from nickel solutions. The relative selectivities of this product for various metals are:

$Fe^{3+} > Zn^{2+} > Al^{3+} > Cu^{2+} > Co^{2+} > Mg^{2+} > Ca^{2+} > Ni^{2+}$

Lewatit® TP272 from Lanxess is a commercial product of this type (Lanxess, 2011a).

Separation processes without chemical regeneration

The following separation processes used in chemical processing applications present the interest that they do not use chemicals for regeneration of the IER and consequently there is no consumption of chemicals and no generation of wastes. The first two cases are based on the Donnan phenomenon, briefly discussed here.

The Donnan equilibrium

When the resin comes in contact with a dilute solution of an electrolyte with the same counter ions as the resin but different co-ions, the concentration of the counter ions in solution is lower than that in the resin while the concentration of the co-ions in solution is higher than that in the resin.

Take a SBA in the Cl^- form as an example (figure 2.4) and in contact with a dilute solution of NaCl. Because of the concentration difference between resin and solution, counter ions (here: Cl^-) will tend to migrate from the resin into the solution phase while co-ions (here : Na^+) will tend to migrate from the solution to the resin phase. However, as the counter ions diffuse out of the resin into the solution and co-ions from the solution into the resin, they create a positive charge in the resin and a negative charge in the solution. The established electric potential between the two phases, called Donnan potential (Helfferich 1962, p. 134), will pull back Cl^- from the solution into the positively charged resin and the Na^+ from the resin into the negatively charged solution. In the end, we will have electroneutrality in both sides but there will be a concentration gradient because there will be more Na+ and Cl- inside the resin relatively to the external concentration than before. This concentration gradient will create an electrical potential gradient which balances the concentration gradient.

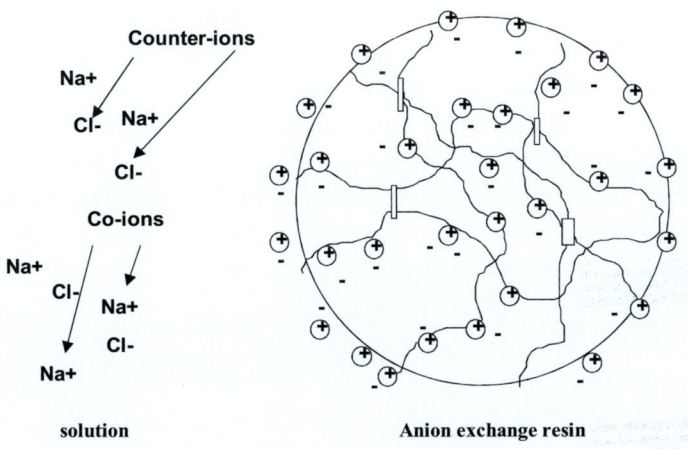

Figure 2.4 Anion exchange resin in contact with a solution of NaCl

The relationship between the electric potential and the concentration gradient is given by the Nernst equation:

$$E = (RT/zF) * \ln(c_{out}/c_{in}) \qquad (2.3)$$

where :
- R=gas const=2 cal mol^{-1}°K^{-1}
- T=temp=273+°C
- z=valence=+1 for Na$^+$ and -1 for Cl$^-$
- F=Faraday const=23000 cal V^{-1}mol^{-1}
- c= concentration in molarity, M

When the electrical potential E_{Na+} for the Na^+ ions will be equal to the electrical potential E_{Cl-} for the Cl^- ions, then there will be an equilibrium, called Donnan equilibrium, where the concentration difference will be balanced by the electrical field.

Take the following example with a resin bead and an external solution of equal volume. R^-N^+ are the functional groups of the strong base anion (SBA) exchange resin and Cl^- are the counterions. Na^+ ions in the external solution are the co-ions. Consider an external solution of 100 mmol/L of NaCl and a resin with 2000 mmol/L of R^-N^+ and 2000 mmol/L of Cl^-. At equilibrium, when the concentration difference will balance the electrical field, some NaCl moved inside the resin beads. We will have, using eq. (2.3),

$$E_{Na+} = (RT/zF) * \ln(c_{out}/c_{in}) \quad \text{for Na+ and}$$
$$E_{Cl-} = (RT/zF) * \ln(c_{out}/c_{in}) \quad \text{for Cl}^-.$$
and at equilibrium $E_{Na+} = E_{Cl-}$.

The concentration of Cl^- ions inside the resin is much higher than in the outside solution, as it is frequently the case.
Trying the concentration of Na^+ inside the resin as 4.6 mmol/L, there will be inside the resin bead 2004.6 mmol/L of Cl^- and 2000 mmol/L of R^-N^+. The concentrations in the external solution will be 95.6 mmol/L of Na+ and Cl^-.
Then c_{out}/c_{in} for Na^+ 95.4/4.6 = 20.739 and
c_{out}/c_{in} for Cl^-: 95.4/2004.6 = 0.04759
and using eq. (2.2) we obtain
$E_{Na+} = 0.07725$ V
$E_{Cl-} = 0.07758$ V

and therefore equilibrium has practically been reached. In this case, only less than 5% of the Na^+Cl^- have moved into the resin.

If the external solution contains $CaCl_2$ instead of NaCl, then $CaCl_2$ is excluded even more, less than 0.24% of $CaCl_2$ has entered the resin beads, as shown below:

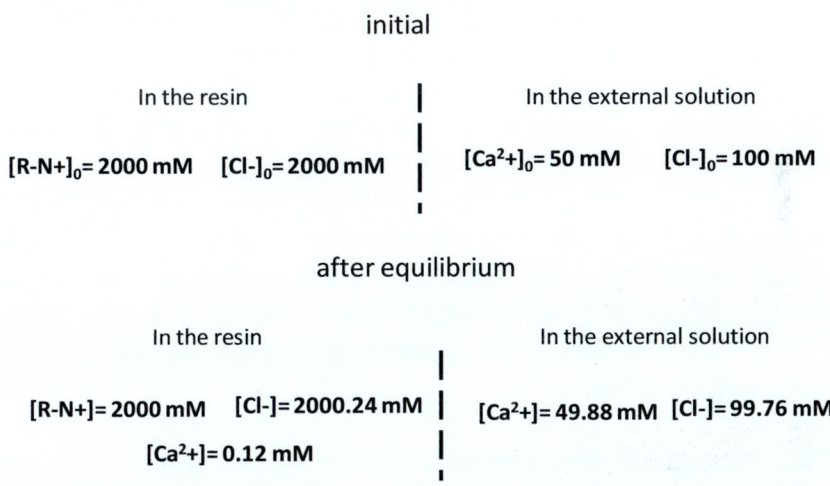

Here we have after the indicated amounts of $CaCl_2$ have moved:

c_{out}/c_{in} for Ca^{2+}: 415.67
c_{out}/c_{in} for Cl^-: 0.0987
and using eq. (2.2),
$E_{Ca^{2+}} = 0.07682$ V
$E_{Cl^-} = 0.07639$ V

Thus, in *an anion exchange resin, the Donnan potential will prevent most of the positively charged co-ions* (the Na^+ or Ca^{2+}) from entering the positively charged (anion exchange) resin and because of electroneutrality, the electrolyte (here NaCl or $CaCl_2$)

will be prevented from entering into the resin. With the concentrations taken in these examples, less than 5% of the NaCl ions or less than 0.24% of $CaCl_2$ in the exterior enter into the resin. This phenomenon is called ion exclusion.

In the same way as it excludes positively charged co-ions, *the resin attracts negatively charged counter-ions,* thus balancing the tendency of the counter-ions to diffuse out of the resin into the solution. *The higher the valency of the counter-ion, the less the electrolyte is excluded by the resin.* For example, the anion exchangers will exclude more NaCl than Na_2SO_4.

Analogous is the situation of *a strong acid cation* (SAC) exchange resin in the Na^+ form in a solution of NaCl. The negatively charged resin *will prevent the co-ions* (Cl^-) *from entering* in the resin and due to the electroneutrality, NaCl will not migrate into the resin.

The greater the concentration difference between the ion exchanger and the solution, the higher is the Donnan potential and the more efficient the ion exclusion. A high total capacity of the resin and a high crosslinking density means high concentration of the counter ion in the resin and for a given counter ion concentration in the solution will tend to increase the Donnan potential and therefore the ion exclusion is increased. *Similarly, the Donnan potential and therefore ion exclusion increases when the solution concentration decreases.*
The Donnan potential required to balance the concentration difference when the counter ion has high valence is smaller because the force acting on the ion is proportional to its valence. Therefore, *ion exclusion is less efficient with counter ions of*

high valency and it is more efficient with counter ions of low valency.

Also, at a given Donnan potential, *co-ions with high valency are more strongly excluded than co-ions of low valency*

For example, a SAC resin will prevent Na_2SO_4 more strongly from entering the resin (here Na^+ is the counter ion and SO_4^{2-} the co-ion) than NaCl (Na^+ is the counter ion and Cl^- the co-ion). The same, a SBA will exclude more $CaCl_2$ than NaCl. On the other hand, a SBA resin will exclude more strongly NaCl (here Na^+ is the co-ion and Cl^- the counter ion) than Na_2SO_4 (Na^+ is the co-ion and SO_4^{2-} the counter ion).

Interactions between counter ions and co-ions can affect the Donnan potential and therefore the ion exclusion. A typical case is the complex formation whereby the Donnan potential is decreased. For example, in a solution of $ZnCl_2$ in contact with a SBA resin in the Cl^- form, the co-ion Zn^{2+} and therefore $ZnCl_2$, is excluded from the resin. However, the formed complex $ZnCl_4^{2-}$ can migrate into the resin.

Ion exclusion

If a SAC resin in A^+ form comes in contact with a dilute solution containing a strong electrolyte A^+Y^- then the cation A^+ will be found at higher concentration inside the resin beads than in the external solution while the anion Y^- (co-ion) will be found in higher concentration in the external solution. There will be then a beginning of migration of cations A^+ from the resin towards the external solution and of anions Y^- from the solution towards the resin. This however will create a net positive charge in the external solution and a net negative charge inside the resin beads building up an electric potential difference between the two phases. This so called Donnan potential will prevent anions Y^- from entering the resin beads and cations A^+ from coming out of the resin. Because of electro neutrality requirement, the co-ion Y^- exclusion from the resin means electrolyte A^+Y^- exclusion. This exclusion is called Donnan exclusion (Helfferich, 1962). From a mixture of electrolytes and nonelectrolytes, the resin excludes the strong electrolytes and takes up the nonelectrolytes. This ion exclusion phenomenon is used to separate chromatographically neutral molecules such as sugar, glycerin etc from mineral salts. The non-electrolyte is eluted from the resin with water. Ion exclusion is used industrially, among others, in molasses desugarization.

Acid retardation

In 1963 a process was introduced, called acid retardation, to separate strong acids from their salts (Hatch and Dillon, 1963). According to this process, the feed solution passes through a strong base anion exchanger which shows a greater affinity for the acid. The acid is retained (retarded) while the salt comes out first. Subsequently, the acid is eluted from the resin with water. This process is based on the Donnan effect and on the fact that the Donnan exclusion is more efficient with co-ions of higher valence and with solutions of low concentration (Helfferich, 1962). With spent pickling solutions for example, or with sulfuric acid anodizing solutions, at high concentrations the acid crosses the Donnan potential and enters the resin while the salts of the dissolved metals, Fe^{2+}, Al^{3+} (the co-ions) are excluded. The major advantage of the acid retardation process is the acid is recovered with a simple water elution.

The process works in two steps:
First, the spent acid passes through the SBA exchange resin from which the de-acidified metal salt solution is collected as by-product. In the second step the retained acid is recovered by stripping the resin with water. Since concentrated solutions are treated, the cycle length is very short, of the order of 1 BV. Therefore, the challenge is to minimize the dead volumes of the installation, to minimize the eluant volume and to minimize mixing of the recovered acid with the metal salts by-product. For that reason, a number of engineering companies have developed systems to handle this type of process, for example the Scanacon Acid Recovery (SAR)® process, the Gütling Retardation KOMParet® system and the Recoflo® Acid Purification Unit APU® process (Eco-Tec Inc) among others.
As an illustration, the Recoflo® (Reciprocating Flow Ion Exchange) APU® process works as follows (Brown, 1997):

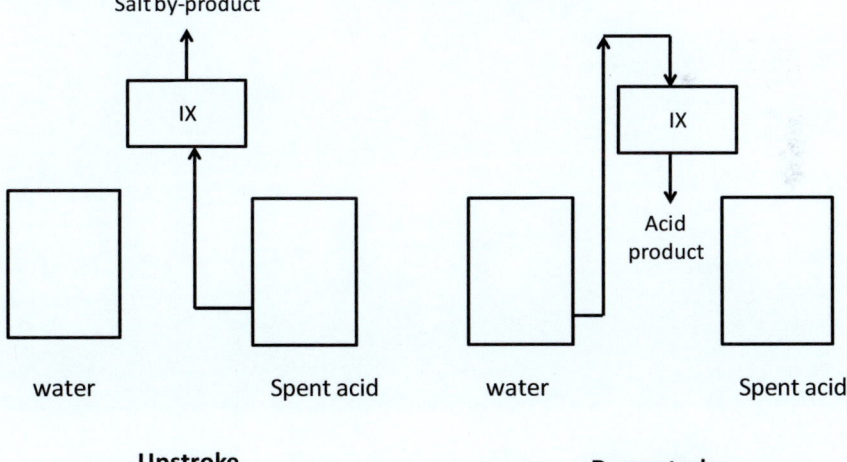

Figure 2.5 Recoflo® APU® process (Eco-Tec Inc.)

The resin, a SBA resin having a small particle size to improve its kinetics, is placed in a compact, short column. In spite of the high pressure drop per meter of resin, the fact that the resin bed is short makes that the total pressure drop is acceptable.
During the first step, called Upstroke, the spent acid is pumped to the bottom of the resin bed. The acid is sorbed by the resin and the remaining salt solution comes out at the top of the resin bed. During the second step, called Downstroke, water is pumped at the top of the resin bed eluting the acid so that purified acid comes out at the bottom of the resin bed. The process is controlled by measuring the stroke volumes. Figure 2.6 illustrates a Recoflo® APU® unit of Eco-Tec, Inc.

Figure 2.6 Recoflo® APU® unit of Eco-Tec, Inc. (Courtesy of Eco-Tec Inc.)

The Recoflo® process has been used, among other applications, in the purification of HCl, H_2SO_4 and mixed acids HNO_3/HF from spent pickle liquors. In case of the mixed acids, the HNO_3 being a strong oxidant, it can oxidize the ion exchange resin to the degree that an explosion can take place. Consequently, care must be taken in this as well as any other application of ion exchange resins where concentrated HNO_3 or other strong oxidants are used, to avoid resin oxidation and possible explosion. In all cases, the operating conditions defined by the process developers must be strictly followed. In the case of mixed acids purification, the process developer recommends low temperatures (<32°C) as well as low HNO_3 concentrations (Brown, 2002).

Ion retardation

In ion retardation, the resin is an amphoteric one having both cation and anion exchange sites. This type of resin can be obtained by polymerizing an anionic monomer, for example acrylic acid, inside a crosslinked anion exchange resin, for example a gel-type styrene-DVB strong base anion exchange resin where the resulting resin is a mixture of linear polyacrylic acid entangled in a styrene-DVB strong base anion network. It can also be obtained by polymerizing a cationic monomer inside a crosslinked cation exchange resin but this type of resin is not commercially available today. The polymerized linear polymer is strongly entangled with the crosslinked resin so that it does not diffuse out. The resulting resin, called snake-cage resin, contains therefore both, cationic as well as anionic groups. In absence of external electrolytes, the cation and anion groups of the resin form an internal salt. When a solution of an electrolyte passes through the resin, the ions from the solution can split the internal salt and be bound on the functional groups of the resin. In this way, the resin absorbs the electrolyte from the solution. The absorbed salt can subsequently be eluted with water. In this way, an electrolyte can be separated from a non-electrolyte, or two electrolytes can be separated from each other, as for example NaCl impurity can be removed from NaOH solution, or separate NaCl from Na_2SO_4, or NH_4Cl from $ZnCl_2$. This resin shows higher selectivity for chloride than for sulfate or hydroxyl ions. In the separation of an electrolyte from a non-electrolyte, the electrolyte comes out after the non-electrolyte while in ion exclusion it is the opposite. The resin can absorb strongly acids but in this case the acid can not be eluted simply with water but it is necessary to wash the resin with NaOH followed by water rinse to bring the resin in the initial condition. Similarly, complete elution of Zn^{2+} in the separation from NH_4Cl was not achieved with water and several cycles were necessary to reach an equilibrium between the incoming and the outgoing Zn^{2+}.

The process is carried out by passing a given volume of the solution of the salt components to be separated through the resin. One of the salts is retarded more than the other, depending on the selectivity of the resin. Then rinse the resin with a few BV of pure water and collect the effluent. A commercial resin of this kind is Dowex® Retardion® 11A8.

This special resin has particular selectivities. For example, it has higher selectivity for sodium chloride than for sodium sulfate. It can therefore separate NaCl from Na_2SO_4 in even concentrated solutions. In some cases, the selectivity for various salts parallels the selectivities of the SBA and the WAC resins which constitute this special retardation resin, for example the resin can separate monovalent from divalent cations such as Zn^{2+} from NH_4^+.

The Recoflo® technology has also been applied in salts separation with a unit named SSU® (Salt Separation Unit), similar to the APU® described above (Brown et al, 1998).

Another amphoteric resin by Mitsubishi Chemical Corp., Diaion® DSR01, where the strong base and the weak acid groups are grafted on the same monomer unit of the polymeric matrix of the resin, is described later (page 200).

Parametric pumping

A parametric pump (Tverdislov *et al*, 1999) consists of an immobile solid phase (the adsorbent) in a jacketed column which alternately retards and releases selected species from a non-adsorbable fluid flowing through the adsorbent. The adsorbent is connected with two reservoirs, one at each end of the column (top and bottom reservoirs). The solution oscillates through the adsorbent giving the opportunity to the components in solution to distribute between liquid and solid phases. By changing a parameter during each oscillation, for example temperature or pH, the equilibrium constant and therefore the distribution of the components changes. For example, considering a feed solution with one component, during downflow operation, the column is cooled and the adsorbent retards the component allowing the solvent to pass. During upflow operation, the column is heated and the adsorbent releases part of the adsorbed compound. Considering the column to consist of a number of theoretical plates (TP), the size of each oscillation is one TP. After a number of oscillations, a steady-state is reached where the top container contains a higher concentration while the bottom container has a lower concentration solution. The difference between the top and bottom concentrations is the separation factor and it depends on the shape of the equilibrium isotherms at the two temperatures. Of course the process presents a practical interest when the two isotherms are significantly different.

Thermal regeneration

The approach consisting of changing a parameter, temperature or electric field for example, to control the adsorption and desorption of compounds or ions on an adsorbent or ion exchange resin was used is various cases. Some years ago (Bolto *et al*, 1976) using an IER containing both weak acid and weak base functional groups on the same copolymer, the regeneration was done at high temperature. The resin was used to partially demineralize brackish waters at low temperatures (ambient) by absorbing salts. Regeneration was done with hot water during which, part of the salts removed during loading were released, due to the lower equilibrium constant at high temperatures compared to low temperatures. The salts concentration in the spent hot water regenerant represented the difference in equilibrium loadings at the two temperatures and was in fact the operating capacity of the resin. This process was called Sirotherm® (Sirotherm® was a Registered Trade Mark of ICI Australia Limited). This interesting process did not have a commercial success however due to the low operating capacity of these special resins and the loss of resin capacity with time due to oxidation.

Electrodeionization

Electrodeionization (EDI) is a process where an IER removes ions from a feed solution, then an electric field is applied across the resin bed that pulls off the charged ions from the resin and directs them to the corresponding oppositely charged electrodes. In this way, the ions are continuously removed from the resin and transferred to an adjacent concentrating compartment. The ions from the concentrate cannot reach the electrodes because they are blocked by a membrane of the same charge. The regenerating H^+ and OH^- ions are either produced electrochemically at the electrodes or by dissociation of the water molecules:
Electrochemical redox reactions:

$2\ H_2O \rightarrow O_2 + 4\ H^+ + 4\ e^-$ (at the anode)
$4\ H_2O + 4\ e^- \rightarrow 4\ OH^- + 2\ H_2$ (at the cathode)

Water dissociation:

$H_2O \leftrightarrows H^+ + OH^-$

The IER is packed between anion permeable membrane (APM) and cation permeable membranes (CPM) or bipolar membrane (BPM). These membranes support physically the IER and direct the charged ions to the right direction. In addition, prevent the concentrate to mix up with the feed solution. IER can be a cation exchanger, an anion exchanger or a mixed bed.

Water regeneration

Some metal cations form anionic complexes in hydrochloric acid media, MCl^{n-}, the speciation of which depends on the Cl^- concentration. For example, Zn^{2+} forms chorocomplexes from Cl^- concentration higher than 0.1 N, Fe^{3+} higher than 1.5 N, Cu^{2+} higher than 3N, UO_2^{2+} higher than 5N and so on for other metals. These anion complexes can then be fixed on an anion exchange resin. After loading, the resin can be regenerated by passing water through, whereby the Cl^- concentration drops below the above mentioned limits and the complexes dissociate to the metal cations, which are released by the resin as Cl^- salts, and the resin remains in the Cl^- form. This can eventually be applied to other complexes. This technique presents an interest not only because water is used as regenerant but also because it allows to separate complex forming metals from other cations that do not.

Another case of water regeneration is illustrated with the removal of H_2SO_4 from solutions using a WBA exchange resin. As long as the concentration is elevated, H_2SO_4 is fixed by the resin as HSO_4^-. By washing the loaded resin with water, the HSO_4^- ions are converted to SO_4^{2-} and H_2SO_4, which exits the column while the SO_4^{2-} ions remain fixed on the resin.

$$2\ R\text{-}NH^+HSO_4^- \xrightarrow{H_2O} (R\text{-}NH^+)_2\ SO_4^{2-} + H_2SO_4$$

An operating capacity of about one half of the total capacity of the resin can be obtained in this way.

Some aspects of ion exchange equilibrium

This section is a reminder of ion exchange equilibrium principles as already discussed in an earlier work (Zaganiaris, 2009). Here this subject is mentioned because it has a direct relevance with certain metal recovery applications discussed in subsequent chapters where metal cations are fixed by ion exchange on a conventional strong acid cation resin from a solution containing also relatively high H^+ concentration. The subject of ion exchange equilibria is treated fully in a number of excellent books on ion exchange (Kunin, 1958; Helfferich, 1962; Arden, 1968).

Let us take an IER containing a given counter ion in contact with a solution containing an electrolyte with different counter ion. In this case, there will be an exchange between the counter ion initially in the resin and the counter ion initially in solution. In general, this exchange is reversible, it can take place when the counter ion in the resin is A and the counter ion in solution is B, or when the counter ion in the resin is B and that in solution A. The extent to which the exchange of the counter ion on the resin for the ions in solution will take place depends upon the preference of the resin for one or another ion. This preference is expressed as a selectivity coefficient $K_{i/j}$ or a separation factor $a_{i/j}$ for a binary exchange. For a multicomponent ion exchange, it is assumed that there are no interferences of one ion on the exchange of another one, so the same selectivity coefficient or separation factor is used.

Consider the case of a SAC resin in the H^+ form that comes in contact with an aqueous solution of NaCl. An exchange takes place of the H^+ for the Na^+. After equilibrium has been established, the IER, initially in the H^+ form, will be found in a form partially H^+ and partially Na^+. Similarly, the solution, initially NaCl, will contain both, NaCl and HCl. The ion exchange reaction is described by eq. (2.4):

$$R\text{-}H + Na^+Cl^- \rightleftarrows R\text{-}Na + H^+Cl^- \quad (2.4)$$

where R represents the polymer matrix of the resin. This equilibrium condition is described with the equilibrium constant (2.5):

$$K = \frac{\{R\text{-}Na\}\{H\}}{\{R\text{-}H\}\{Na\}} \quad (2.5)$$

where K is the equilibrium constant and {} denote the activities of H^+ and Na^+ on the resin and in solution after equilibrium. For simplicity, if we use concentrations instead of activities, eq. (2.2) becomes:

$$K_{Na/H} = \frac{(R\text{-}Na)(H)}{(R\text{-}H)(Na)} = \frac{q_{Na}\, c_H}{q_H\, c_{Na}} \quad (2.6)$$

where $K_{Na/H}$ is the selectivity coefficient and ( ) denote concentrations in equivalents per liter (eq/L) or equivalents per liter resin (eq/L_R).

q_{Na} = concentration of Na in resin in eq/L_R
c_H = concentration of H in solution in eq/L
q_H = concentration of H in resin in eq/L_R
c_{Na} = concentration of Na in solution in eq/L

Alternatively, the use of the separation factor, defined as:

$$\alpha_{i/j} = \frac{\text{distribution of ion i between phases}}{\text{distribution of ion j between phases}} = \frac{y_i / x_i}{y_j / x_j} \quad (2.7)$$

is more practical. Here :
  x_i is the equivalent fraction of ion i in solution and
  y_i is the equivalent fraction of ion i in the resin.

$$x_i = c_i / C \quad \text{and} \quad y_i = q_i / q \quad (2.8)$$

Where c_i is the concentration of ion i in solution in eq/L and C the total concentration of all ions in solution in eq/L. q_i is the concentration of ion i on the resin in eq/L_R and q is the total capacity of the resin in eq/L_R.
Using $x_j=1-x_i$ and $y_j=1-y_i$, eq 2.7 becomes:

$$y_i = \alpha_{i/j} * x_i / (1+(\alpha_{i/j} -1)*x_i) \quad (2.8a)$$

If the resin prefers ion i over ion j we call this a favorable equilibrium and $\alpha_{i/j} > 1$. In the opposite case we call this an unfavorable equilibrium and $\alpha_{i/j} < 1$.

Using eqs (2.4) and (2.5) we have for the above example of the H^+/Na^+ exchange :

$$\alpha_{Na/H} = \frac{q_{Na}\, c_H}{q_H\, c_{Na}} = K_{Na/H} \quad (2.9)$$

From a plot of the equilibrium concentration of one of the ions in the resin versus the equilibrium concentration of that ion in solution, at constant temperature, one can obtain the value of $\alpha_{Na/H}$ or of $K_{Na/H}$. This type of plots is called ion exchange isotherms.

Figure 2.6 illustrates the plot of y_i vs x_i for a mono-monovalent equilibrium. The curve that gives a separation factor of 3, which gives a curve convex to the x-axis, represents a favorable equilibrium. The curve which is concave to the x-axis and gives a separation factor of 0.5 is an unfavorable equilibrium.

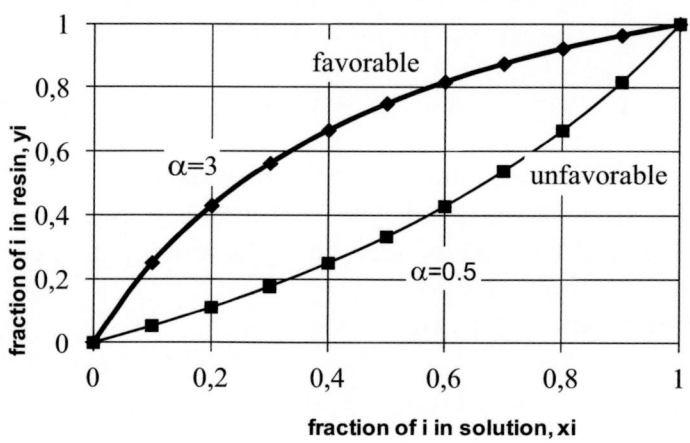

Figure 2.7 Isotherms for favorable and unfavorable equilibrium

For the general case:

$$a A + b R\text{-}B \leftrightarrows a R\text{-}A + bB \quad (2.10)$$

we have:

$$K_{A/B} = \frac{q_A^a * c_B^b}{q_B^b * c_A^a} \quad (2.11)$$

and

$$\alpha_{A/B} = \frac{y_A * x_B}{y_B * x_A} = \frac{q_A * c_B}{q_B * c_A} \quad (2.12)$$

From (2.11) and (2.12) it results:

$$(\alpha_{A/B})^b = K_{A/B} * (q_A/c_A)^{b-a} \quad \text{or} \quad (\alpha_{A/B})^a = K_{A/B}*(q_B/c_B)^{b-a} \tag{2.13}$$

Therefore, the separation factor is not constant but it depends on the composition of the solution and the resin for the case where a is different from b. If we consider the case where ion A in eq. (2.8) is divalent and ion B is monovalent and therefore a=1 and b=2, and using fractions rather than concentrations (eq. (2.8)) then eq. (2.11) becomes:

$$\frac{y_A}{(1-y_A)^2} = K_{A/B} \frac{q}{C} \frac{x_A}{(1-x_A)^2} \tag{2.14}$$

where q is the total capacity of the resin in eq/L_R and C the total ionic concentration of the ions in the solution, in eq/L.

From eq (2.14) it is seen that for ions of different valence, the isotherm would depend on the total ionic concentration in solution, since the total resin capacity, q, is constant.

In multicomponent systems, one can establish a selectivity sequence by comparing the separation factors, $\alpha_{A/B}$, or the selectivity coefficients, $K_{A/B}$, of various ions with respect to one ion taken as a reference. As indicated above, selectivity sequence based on separation factors depend on the total ion concentration in solution.

There are some rules that characterize the relative selectivities of a resin for different ions:
- selectivity increases as the size of the (hydrated) ion decreases

- selectivity increases as the valence of the ions increases
- selectivity increases as the degree of crosslinking of the resin increases

In the table below they are indicated as an illustration, the selectivities of SAC resins having different DVB content for various cations, based on selectivity coefficients (de Dardel and Arden, 1989) :

TABLE 2.1 Selectivity coefficients of SAC resins

Cation	degree of crosslinking (% DVB)			
	4	8	12	16
H^+	1	1	1	1
Na^+	1.3	1.5	1.7	1.9
K^+	1.75	2.5	3.05	3.35
Zn^{2+}	2.6	2.7	2.8	3.0
Cu^{2+}	2.7	2.9	3.1	3.6
Ni^{2+}	2.85	3.0	3.1	3.25
Ca^{2+}	3.4	3.9	4.6	5.8
Pb^{2+}	5.4	7.5	10.1	14.5

The above principles have at least two implications in applications of IER in chemical processing, in particular in metals removal. The first concerns the removal of metal ions from acidic solutions with a SAC resin in the H^+ form. The second concerns the regeneration of SAC resins used in metals removal.

As an illustration for the first case, we consider a SAC resin in the H^+ form and a solution containing an acid and a divalent metal, M^{2+}, present as cation. In order to remove as much as possible of the metal from the acidic solution, a resin should be chosen that has the maximum possible selectivity coefficient for the metal with respect to H^+. From table 2.1 this means to choose a high DVB content resin. Assume that the selectivity coefficient $K_{M/H}$ is 3, which is a representative number for a number of common metals at high DVB content (Table 2.1). Figure 2.8 illustrates the equilibrium isotherms of the exchange:

$$2\ R\text{-}SO_3^-\ H^+ + M^{2+} \leftrightarrows (R\text{-}SO_3^-)_2\ M^{2+} + 2\ H^+ \quad (2.15)$$

for two values of the total ion concentration, C, in solution. As it is observed, the curve at low total concentration is a favorable equilibrium (curve convex to the x-axis) while at high total concentration is an unfavorable equilibrium (curve concave to the x-axis).

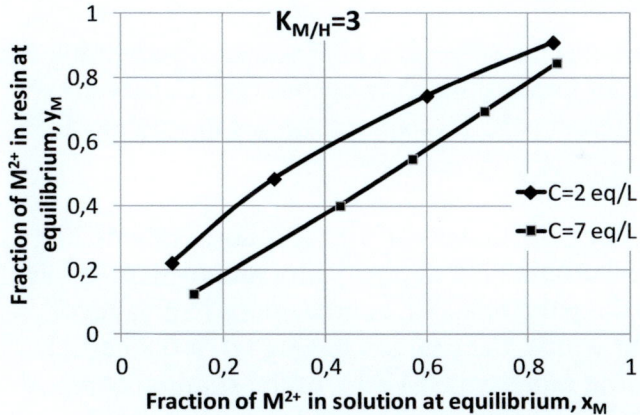

Figure 2.8 Equilibrium isotherms for the exchange
$2\ R\text{-}SO_3^-\ H^+ + M^{2+} \leftrightarrows (R\text{-}SO_3^-)_2\ M^{2+} + 2\ H^+$

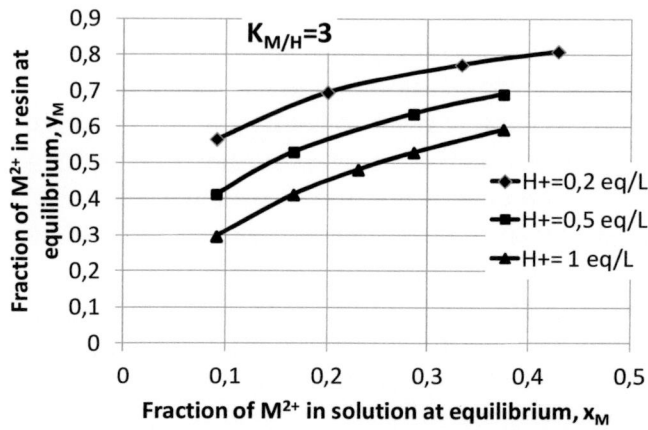

Figure 2.9 equilibrium isotherms for the exchange
$2\ R\text{-}SO_3^-\ H^+\ +\ M^{2+}\ \leftrightarrows\ (R\text{-}SO_3^-)_2\ M^{2+}\ +\ 2\ H^+$
at different H^+ concentrations at equilibrium

Figure 2.9 illustrates the same data but instead of presenting the curves at different total solution concentrations, C, they are presented at different acid, H^+, concentrations in solution at equilibrium.

For example, taking the curve of $(H^+) = 0.2$ eq/L, (about 20 g H_2SO_4/L taking sulfuric acid as practically monovalent at this concentration) the point at $x_M=0.2$ corresponds to a value of $y_M=0.7$. In other words, the solution having $(H^+)=0.2$ eq/L, the (H^+) concentration represents the 80% of the solution composition (since $x_M=0.2$). Therefore, the total ionic concentration in solution will be $100*0.2/80= 0.25$ eq/L. The metal ion concentration will then be 0.05 eq/L. The saturation capacity with this solution composition is 70% of the total resin capacity. Taking the total resin capacity as 2 eq/L_R, the saturation capacity is expected to be about 1.4 eq/L_R. The saturation capacity is the resin capacity obtained in a column run when the effluent reaches the same composition as the influent. The operating capacity, which is of practical interest, is the capacity at a given breakthrough end-point. The ratio of operating capacity to saturation capacity depends on the slope of the breakthrough curve (2.10).

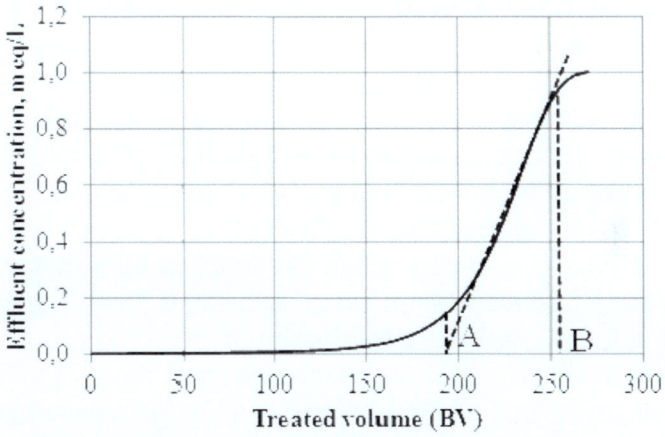

Figure 2.10 Breakthrough curve

The more favorable is the equilibrium isotherm, the higher will be this slope. At AB = 0 the operating capacity is equal to the saturation capacity. If 0A = AB then the operating capacity is 2/3 of the saturation capacity. These apply when the exchange sites of the resin are all, or most, in the regenerated form. If after regeneration, a significant part of the exchange sites of the resin remain in the exhausted form, then the breakthrough, point A in figure 2.10, and the saturation, point B, take place earlier.

In this numerical example given above, the operating capacity can be assumed to have a high fraction of the saturation capacity (low ionic concentrations, favorable equilibrium), indicating that a SAC resin can be envisaged to purify the solution from the metal ions.

On the other hand, at a H_2SO_4 concentration of 50 g/L (about 0.5 eq/L taking H_2SO_4 as monovalent at this concentration) and 0.05 eq/L of the metal concentration as in the previous example, the total solution concentration is 0.55 eq/L and $x_M = 0.05/0.55 = 0.09$. Then y_M is 0.45 and the saturation capacity would be $0.45*2=0.9$ eq/L_R. In this case, the equilibrium is less favorable than in the previous example (because of a higher total ionic concentration) and therefore it is expected that the operating capacity would be lower. This would possibly be a rather low capacity to envisage a SAC in this case.

If in this case one dilutes down the solution by a factor of two, where the H^+ concentration would become 0.25 eq/L and that of M^{2+} 0.025 eq/L, giving a total solution concentration of 0.275 eq/L and again a fraction $x_M=0.09$ then from figure 2.8, the saturation capacity is $0.53*2=1.06$ eq/L_R which is somewhat higher than the previously found 0.9 eq/L_R. The positive effect of the dilution on the resin capacity originates from the fact that the equilibrium isotherm in the case of mono-divalent equilibrium depends on the total solution concentration (fig. 2.8). If we have had a mono-monovalent equilibrium, diluting down would have no effect on the resin capacity since the equilibrium isotherm does not depend on the solution concentration (fig. 2.7).

Since a high DVB resin was preferred in order to have a high selectivity coefficient for the metal cations with respect to the H^+, the regeneration of the exhausted resin, which is the reverse reaction of eq 2.15, should be very unfavorable. A high regeneration level is therefore required in these applications. In general, levels of 100-250 g HCl/L_R are practiced, depending on the metal and the DVB content of the resin. If H_2SO_4 is used as regenerant, these levels become 150-300 g H_2SO_4/L_R. Since the equilibrium curves depend on the total solution concentration, fig. 2.8, where the monovalent H^+ is preferred by the resin at high total solution concentrations, a high concentration of the acid regenerant helps.

3. Metals removal or recovery with IX

In this section they are put together all applications where some metal is removed from a solution using IER. It includes metal processes where metals constitute, at least partially, the final product, and other industries where the final product is not a metal but in the process, some metals are present as impurities or as catalysts or making part of the process itself, which are removed by ion exchange and eventually recycled back to the process.

In solution, the metals can be found either as cations or as (anionic) complexes. In the form of a cation, they can be removed either with a cation exchanger, a strong acid cation or a weak acid cation, or a selective resin. When the metal cations to be removed are found in a low salt background or in only slightly acidic pH, the use of a SAC resin in the H^+ form to remove the metals, along with the rest of the cations, is straight forward. Similarly a carboxylic resin can be used to remove heavy metals due to the high selectivity for these cations. This is the case for example of the recycling of plating rinse waters containing no more than a few hundreds of ppm's of various salts including metals where SAC and WBA or SBA are used to demineralize them. On the other hand, there are cases where it is desired to remove metal impurities, for example from plating or other baths, where there are some grams per liter of metals in a high acid background solution. In that case, it is not immediately obvious if a SAC in the H^+ form would be capable to remove economically the metals from such a solution. In those cases one should be able, by using equilibrium principles and the relative affinities of the resin for various cations, to estimate the composition of the resin at equilibrium with a solution having the composition of the bath in question. This equilibrium resin composition will help to decide whether this resin might be suitable in purifying the bath from the metals. These considerations were discussed in the previous chapter. In the cases where metals are found in high salt background, or only certain metals need to be removed from a solution containing other metals as well, then a selective resin is appropriate. Here carboxylic resins form too weak complexes compared to selective resins discussed in the previous chapter to be employed.

Metal processing

In the metal processing industry, several operations are performed in order to give to the metals certain properties such as corrosion resistance, hardness, improved painting, decoration etc. These operations include *degreasing, etching, pickling, chemical passivation, anodizing, chemical mechanical polishing (CMP), conversion coating, electroplating, electroless plating* and *waste waters treatment* to remove the last traces of heavy metals before the effluents are discharged. In these operations, a solution of an electrolyte reacts with the metal whereby the concentration of the active electrolyte decreases while the concentration of impurities increases. While the active ingredient can be replenished by adding more, the concentration of the impurities continues to grow. As the concentration of the impurities goes up, the quality of the desired treatment on the metal deteriorates. To face this situation, either the contaminated solution is discarded, or by using an appropriate treatment, the impurities can be removed so that the life time of the electrolyte solution is extended. Ion exchange technology can make significant contribution in chemicals recovery and reduction of wastes to treat.

The main applications of IX in metal processing are:
- raw water softening or deionization to produce high quality process water;
- rinse waters purification and water recycling eventually with some metal recovery;
- plating baths purification;
- waste waters purification.

In general, rinsing of the plated items is done in two steps. In the first, a minimum quantity of water is used and the rinse waters return to the plating bath. In the second step, large quantities of water are used. These waters are then treated with conventional SAC and SBA or WBA exchange in order to recycle back to rinse, or they are treated with one selective resin in order to recover some metal, as is the case in figure 3.1. In this case, the effluent is not completely deionized and goes for further treatment.

Alternative technologies to remove metals from water and wastewaters include: membrane filtration, solvent extraction, electrolysis (electrowinning), adsorption (activated carbon, hydrous metal oxides, various low-cost adsorbents), chemical precipitation, coagulation or the addition of ligands in the solution.

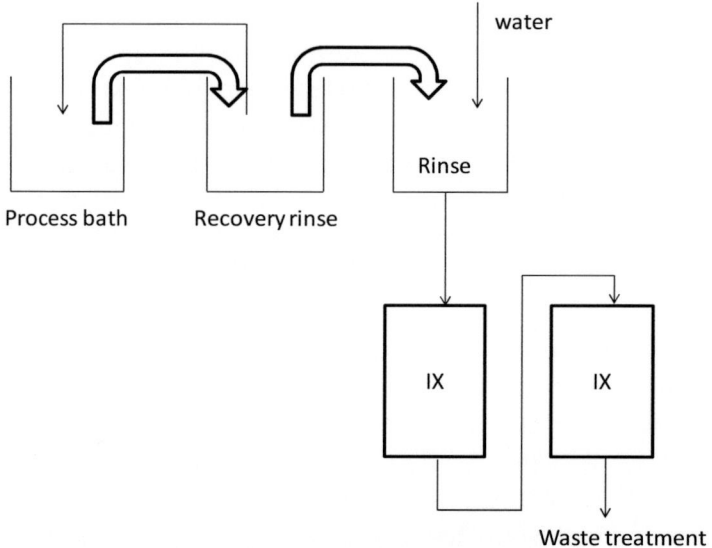

Figure 3.1. Cu recovery from rinse waters

In what follows, it is reviewed the various applications of ion exchange in the metal processing industry except the water softening or demineralization.

It should be pointed out that on a given metal part, more than one treatment can be applied. For example, a part to be chromium plated, may undergo the following stages:
- cleaning to remove dirt and surface impurities
- pickling
- Copper cyanide plating
- Nickel sulfate plating
- Chromium plating

A zinc plating is applied on steel to protect from red rust while a Cr^{3+} passivating coating is subsequently applied to protect the zinc coating from white rust. Therefore at a metal finishing plant it may be various ion exchange treatments on single or combined solutions.

The main metals that are plated are chromium, copper, cadmium, zinc, nickel, silver and gold. Some of the baths are acidic while some are alkaline, where the metals are found as cyanide complexes.

Pickling liquors purification

In metal plating or galvanizing industries, the preparation of the metal surfaces to remove metal oxides is performed with strong mineral acids. HCl or H_2SO_4 are mostly used for carbon steel while mixtures of HNO_3 and HF are used for stainless steel.

HCl

When HCl is used, the initial HCl concentration is between 16 and 18% w/w. The acid is used continuously until the concentration drops to about 2-5% where further use of the spent HCl becomes uneconomical and unpractical due to the slow reaction. At this point, the spent HCl containing about 80-120 g Fe^{2+}/L, can be recovered from the spent pickling liquor using different techniques such as evaporation, pyrohydrolysis or hydrothermal regeneration.

Pyrohydrolysis takes place according to the reactions:

$$4\ FeCl_2 + 4\ H_2O + O_2 \longrightarrow 8\ HCl + 2\ Fe_2O_3$$
$$2\ FeCl_3 + 3\ H_2O \longrightarrow 6\ HCl + Fe_2O_3$$

Hydrothermal regeneration takes place according to the reactions:

$$12\ FeCl_2 + 3\ O_2 \longrightarrow 8\ FeCl_3 + 2\ Fe_2O_3 \quad \text{(oxidation)}$$
$$2\ FeCl_3 + 2\ H_2O \longrightarrow 6\ HCl + Fe_2O_3 \quad \text{(hydrolysis)}$$

In all cases, iron is converted to iron oxide powder and HCl is recovered in the recuperator at 18% strength.

In galvanizing operations, the spent HCl contains in addition to Fe^{2+}, Zn^{2+} coming from stripping Zn^{2+} from rejected articles. The Zn^{2+} concentration in the spent HCl can be in the range of 20 g Zn^{2+}/L. The presence of Zn^{2+} in the spent pickling solution causes process difficulties in the HCl recovery and therefore it is desirable to remove it before recovering HCl.

Zn^{2+} in 0.5M HCl or higher is found as anionic complex $ZnCl_n^{2-n}$

where n=2, 3 or 4:

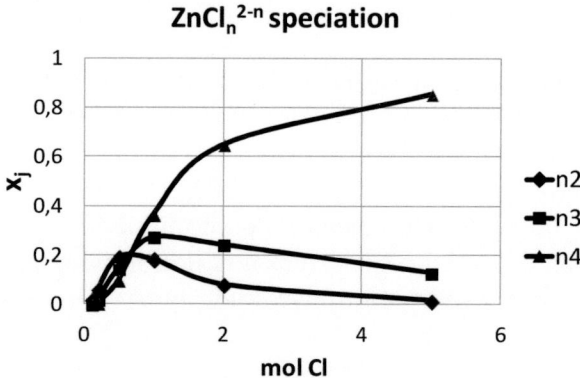

Figure 3.2 $ZnCl_n^{2-n}$ speciation in Cl⁻ media

Therefore, in the spent pickle liquor (HCl is 1-2 M), Zn is found predominately as $ZnCl_4^{2-}$ and some $ZnCl_3^-$. Under these conditions, Fe^{2+} does not form anionic complexes with HCl and therefore Zn can be selectively removed with an anion exchange resin (Haines et al, 1973; Bäcklund and Rennerfelt, 1981), in spite of the high concentration of Cl^- in the solution. In fact, Zn^{2+} forms the $ZnCl_4^{2-}$ complex which is not excluded by the Donnan effect and can then enter the resin where it is fixed by exchange with the counter-ion (Helfferich 1962) or even species such as $ZnCl_3^-$ or $ZnCl_2$ can form the higher complex $ZnCl_4^{2-}$ on the functional group of the resin.

The ion exchange reaction during the loading step is:

$$2\,R^+Cl^- + ZnCl_4^{2-} \leftrightarrows (R^+)_2\,ZnCl_4^{2-} + 2\,Cl^-$$

Both strong and weak base resins have been tried and both can be used. Due to the relatively high Zn concentration in the liquor (it can be about 15 g Zn/L), the treated volume per cycle is about 2-3 BV (operating capacity can be 30-40 g Zn/L_R or about 0.9-1.2 eq/L_R). Therefore, the specific flow rate is in the range 1-2 BV/h so that the service cycle is around 1-2 hours. Regeneration is done with 2-5 BV of water during which, the decrease in the Cl^- concentration in solution results in the formation of $ZnCl_2$ which leaves the resin leaving the functional groups in the Cl^- form, thus the resin being ready for the following cycle:

$$(R^+)_2ZnCl_4^{2-} \leftrightarrows 2R^+Cl^- + ZnCl_2$$

One thing to comment here is the fact that the liquor density is in the range of 1.2 g/cc while the anion exchange resin density in water, thus at the end of regeneration, is in the range of 1.1

thus raising the question of resin floating. However, soon after the beginning of the loading cycle, the resin will be found impregnated with the pickle liquor and its density will increase to reach or to exceed, depending on the skeletal density of the resin, that of the pickle liquor.

The flow diagram for Zn removal from spent HCl pickle liquors is outlined in figure 3.3.

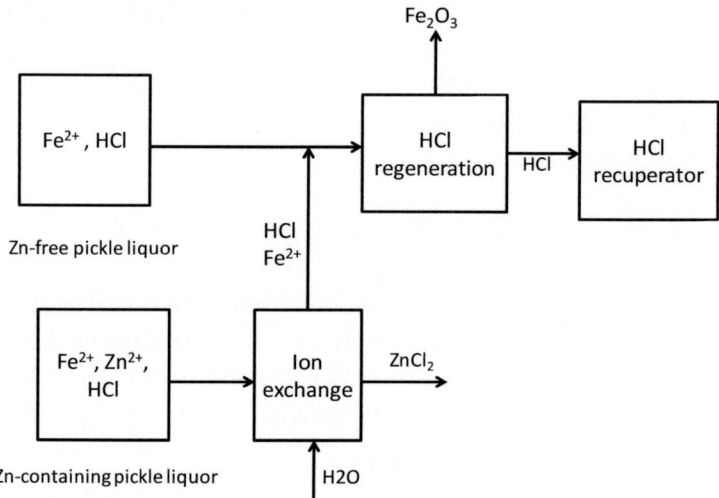

Figure 3.3   Zn removal from spent HCl pickle liquors

Rinse waters from HCl pickling have been treated with a SAC resin to remove Fe and Zn^{2+}, both being as cations because the concentration of Cl^- was too low to form anionic complexes (Maranon et al, 2005). Concentrations of Fe and Zn were about 300 ppm and 200 ppm respectively.

H₂SO₄

Fresh H_2SO_4 pickle liquor contains about 15 % w/w H_2SO_4 while the spent acid contains typically 8-10 % H_2SO_4 w/w and 7-8 % $FeSO_4$ (typically 110 g H_2SO_4/L and 70 g $FeSO_4$/L). H_2SO_4 is recovered from the spent liquor by low temperature crystallization of $FeSO_4$.

An ion exchange method has been described to separate the H_2SO_4 from $FeSO_4$ and recycle the H_2SO_4 using a strong acid cation exchanger (Fradkin and Tooper, 1955). The reaction during loading is:

$$2\,R^-H^+ + FeSO_4 \leftrightarrows (R^-)_2\,Fe^{2+} + H_2SO_4$$

thus recovering the H_2SO_4 bound on the Fe^{2+} and recycling it together with the free H_2SO_4 of the spent liquor.

Regeneration was done with either H_2SO_4 or HCl with preference to HCl which allows the recovery of HCl from the spent regenerant. The Fe salts are collected from the eluates as by-products. In the referenced work, with a feed solution containing 140 g H_2SO_4/L and 140 g $FeSO_4$/L, an operating capacity of 0.7 e/L_R of $FeSO_4$ was obtained using 190 g HCl/L_R regeneration level. The resin was Nalcite HCR, a gel-type SAC.

Mixed HNO₃/HF

Pickling of stainless steel uses mixed HNO_3 and HF acids. Fresh pickling solution contains 90-160 g/L HNO_3 and 10-40 g/L HF. HNO_3 being a strong oxidant oxidizes the metal oxides and metals to metal cations like Fe^{3+}, Cr^{3+} and Ni^{2+} which then form complexes with HF:

$$4Fe + 8HF + 4HNO_3 \longrightarrow 6\,HNO_2 + 4FeF_2^+ + 4NO_3^- + 6H_2O$$

The mixture HNO_3 and HF acids can be recovered using the spray roasting technique at high temperatures (400°C) where metal fluorides and nitrates are converted to the oxides while the acids are sent to an absorption column and to acid recovery tank. Some H_2O_2 is added to the absorber in order to oxidize any NO and NO_2 gases formed during the decomposition of the metal fluorides and nitrates, to HNO_3.

Opposed to this technique, ion exchange process to recover mixed acids pickling baths has been developed (Brown, 2002) using the acid retardation process, as described previously (p. 70). In purifying the mixed acids pickling solution, the mixture of HNO_3/HF acids containing metal salts passes through a SBA resin which attracts the acids more, thus retarding the exit of the acids, letting the metal salts come out first. The metal salts go then to the wastes while the free HNO_3 and HF acids are eluted from the resin with water and return to the pickling bath. In figure 3.4 it is illustrated a Recoflo® APU® unit of Eco-Tec, Inc. for pickle liquors purification in a stainless steel plant.

Figure 3.4 Eco-Tec APU® unit for purifying pickle liquors in stainless steel plant (Courtesy of Eco-Tec Inc.)

HNO_3 acid being a strong oxidizing agent however, can oxidize the ion exchange resin. In order to avoid resin oxidation with a risk of violent reaction causing explosion, the resin should be cooled below 32°C (Brown, 2002) and the concentration of nitric acid in the solution should be kept low.

Another technique used to recover rinse waters of mixed acids pickling of stainless steel is a combined RO-IX technique (Schmidt et al, 2005). Typical used rinse waters composition is 1 g Fe^{3+}/L, 0.14 g Cr^{3+}/L, 0.07 g Ni^{2+}/L, 2 g HNO_3/L and 1 g HF/L. The concentrate of the RO is sent to the regeneration of the pickling solution. The permeate of the RO containing about 0.4-0.5 g/L of HNO_3 and HF, and less than 10 mg/L of metals was polished with an acrylic WBA resin to remove the residual acids in order to make it suitable for rinsing, as shown in the figure below:

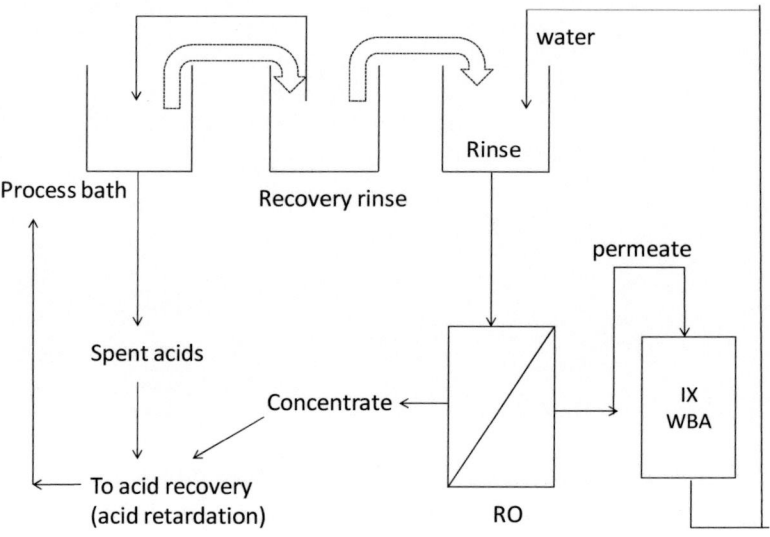

Figure 3.5 Stainless steel pickling with mixed acids combining IX and RO (Schmidt et al, 2005)

Chemical passivation

Passivation is the formation of a coating, usually an oxide, on the surtace of a metal in order to make it more resistant against corrosion. Passivating treatments include: chromic acid, Cr^{3+} salts, nitric acid, phosphoric acid, citric acid.

Cr^{3+} passivation of zinc surfaces is a conversion coating converting zinc to a coating comprising zinc and Cr^{3+} salts. Purification of the Cr^{3+} solutions from Zn^{2+} and Fe^{3+} impurities is achieved with a selective ion exchange resin of the BPA type. Another option is a Diphonix® type resin such as Purolite® S-957 (Fernandez-Olmo, 2008) which can be used to purify Cr^{3+} solutions from Fe^{3+}. Lewatit VO-OC 1026, a D2EHPA impregnated resin, can also be used for purification of the Cr^{3+} solutions from Zn^{2+} and Fe^{3+} impurities. Regeneration is done with 15% HCl or 15% H_2SO_4.

Phosphoric acid baths in passivating iron or steel contain a phosphate salt of manganese or zinc or iron dissolved in H_3PO_4. Upon contact with the steel, there is a reaction of H_3PO_4 with the steel giving off H_2:

$$Fe° + 2\ H_3PO_4 \longrightarrow Fe(H_2PO_4)_2 + H_2 \uparrow$$

This reaction leads to a depletion of free acid and raises locally the pH so that the tertiary phosphate salt precipitates on the surface of the steel. In the case of Zn phosphating bath the reaction is,

$$Zn(H_2PO_4)_2 \leftrightarrows ZnHPO_4 + H_3PO_4$$

$$3\ ZnHPO_4 \leftrightarrows Zn_3(PO_4)_2 \downarrow + H_3PO_4$$

Spent phosphoric acid bath can be purified and recycled using a SAC resin in the H^+ form to remove Fe^{2+} and other metals that accumulate in the bath. The resin can remove about 35 g Fe^{2+}/L_R. Regeneration is done with 2 BV of 10-15% HCl or 10-15% H_2SO_4.

Rinse waters from phosphoric acid passivation can be treated with a SAC resin in the H^+ form to remove cationic impurities such as Zn^{2+}, Fe^{2+} and Mn^{2+}. After the SAC, a WBA follows that removes the PO_4^{3-}. If the waters contain some detergent, then it is advisable to have a scavenger to fix the detergent, either activated carbon or an adsorbent. Otherwise, the detergents risk to foul the resins.

Aluminium anodizing

Anodizing is an electrochemical process which converts the surface of a metal to the oxide, essentially to give a corrosion resistance to the metal. Aluminium anodizing is the most frequent, although other metals can also be anodized.
Anodizing consists in immersing aluminium in an acid bath and allowing an electric current to pass through. The metal to anodize is the anode (hence the term anodizing) where oxygen is released and forms the metal oxide. The overall reaction in both electrodes is:

$$2\ Al + 3\ H_2O \longrightarrow Al_2O_3 + 3\ H$$

Before anodizing aluminum, there is a bright dip finishing step using 65-80% H_3PO_4 and other ingredients. Used bath contains some 30-45 g/L of aluminum, as Al. Most of this acid is drugged out during rinse. The recovery rinse waters are recirculated so that the H_3PO_4 concentration becomes 10-20%. At this concentration of H_3PO_4 it is possible to purify H_3PO_4 by removing Al^{3+} and other cations with a SAC resin in the H^+ form using the Eco-Tec Recoflo® process (Munns and Sullivan, 1995). Regeneration of the resin is achieved with H_2SO_4. The excess H_2SO_4 is recovered using the acid retardation process described previously (page 70).

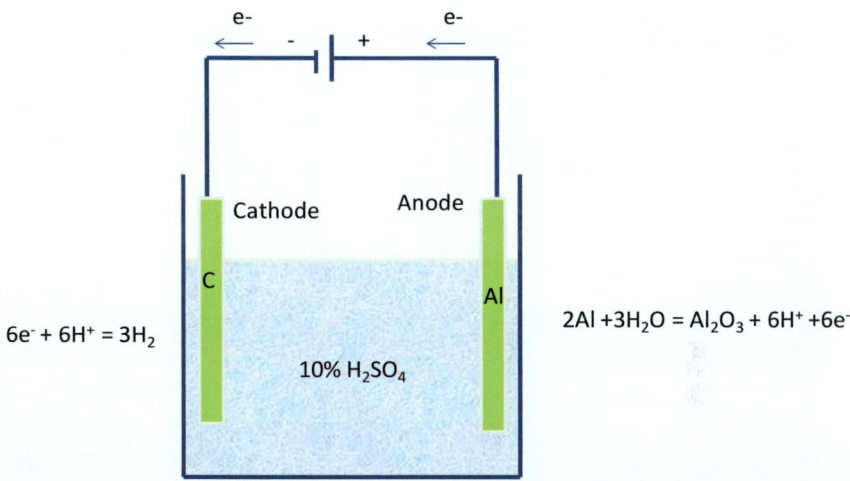

Figure 3.6 Aluminium anodizing

In aluminum anodizing, the acid solutions used are chromic acid or sulfuric acid which is the more widely used today. The H_2SO_4 in the fresh anodizing bath has a concentration of 10%. When aluminum reaches the level of 15-20 g/L the solution is decanted. The recovery of H_2SO_4 from the spent anodizing solution is ideally achieved with the acid retardation process described above. By passing the spent solution through a strong base anion exchanger in upflow direction, the aluminum salt comes out at the top of the column while the acid is retained. By passing water from the top of the column, purified acid is collected at the bottom. This is illustrated in figure 3.7 which shows a purification of an aluminum anodizing liquor by Eco-Tec Inc. called AnoPur^{TM}. The two vessels on either side are the acid tank and the water tank. In the middle is the resin column.

Figure 3.7 AnoPur^{TM} system purifying acid for aluminum anodizing plant. (Courtesy of Eco-Tec Inc.)

Fresh chromic acid in anodizing baths contains about 100 g/L CrO_3. When Al^{3+} reaches the level of 12 g/L, the solution is discarded. The purification of chromic acid anodizing bath is achieved with a macroreticular strong acid cation exchange resin in the H^+ form. Because the concentration of chromic acid is not very elevated, the solution is treated directly on the SAC resin. This resin can treat about 4 BV of such a solution with less than 1 g Al_2O_3/L average leakage. The flow rate is about 3 BV/h and regeneration is performed with 10-15% HCl at a level of 300 g HCl/L_R.

Metal plating

In metal plating, a metal is deposited on a surface in order to give to this surface corrosion resistance, increased hardness, improved paint adhesion or other properties. Jewelry uses plating to give gold or silver finish. There are several plating techniques, electroplating being the most frequent.

Electroplating

The principle of electroplating is illustrated in the figure 3.8 below. The plating bath contains a salt, here $CuSO_4$, of the metal to be applied, here Cu. The object to be plated, here the metal M, is connected to the cathode of a bettery or rectifier (negative pole). The anode is made of the metal to be applied, here Cu. The half reaction at the cathode (reduction, gain of electrons), the more positive is the $E°red$, the more tendency it has to take place. At the anode (oxidation, loss of electrons) the more negative is the $E°_{ox}$ the more difficult is for the half reaction to take place.

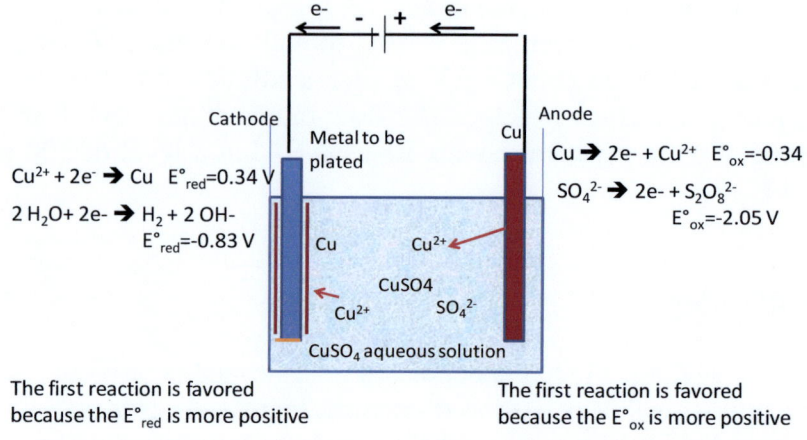

Figure 3.8 Electroplating a metal M with copper in a CuSO$_4$ bath.

Some standard reduction potentials are given in the table below:
Table 3.1 Standard reduction potentials, E°$_{red}$

Acidic solutions	E°$_{red}$ (V)
$S_2O_8^{2-} + 2e^- \longrightarrow 2\ SO_4^{2-}$	+2.01
$Au^+ + e^- \longrightarrow Au$	+1.68
$Cl_2 + 2e^- \longrightarrow 2\ Cl^-$	+1.36
$Cr_2O_7^{2-} + 14\ H^+ + 6e^- \longrightarrow 2\ Cr^{3+} + 7H_2O$	+1.33
$O_2 + 4H^+ + 4e^- \longrightarrow 2\ H_2O$	+1.23
$Cu^{2+} + 2e^- \longrightarrow Cu$	+0.337
$2\ H^+ + 2e^- \longrightarrow H_2$ (reference)	0
$Ni^{2+} + 2e^- \longrightarrow Ni$	-0.25
$Cr^{3+} + 3e^- \longrightarrow Cr$	-0.74
$Zn^{2+} + 2e^- \longrightarrow Zn$	-0.763
$Al^{3+} + 3e^- \longrightarrow Al$	-1.66
$Na^+ + e^- \longrightarrow Na$	-2.714

Alkaline solutions	
$O_2 + 2H_2O + 4e^- \longrightarrow 4\ OH^-$	+0.4
$2\ H_2O + 2e^- \longrightarrow H_2 + 2\ OH^-$	-0.83

In the figure 3.8 when current passes, the following half reactions take place:

Cathode:
$Cu^{2+} + 2e^- \longrightarrow 2\ Cu°$ $E°_{red} = +0.337$ V
Anode:
$2Cu° \longrightarrow 2\ Cu^{2+} + 2\ e^-$ $E°_{ox} = -0.337$ V
Another possible reaction that could have taken place at the anode is:
$SO_4^{2-} \longrightarrow S_2O_8^{2-} + 2e^-$ $E°_{ox} = -2.01$ V
Since however the $E°_{ox}$ for this reaction is more negative, it is the oxidation of $Cu°$ to Cu^{2+} that will take place.

It is possible that the anode is made of a non-consumable material, such as lead. In that case, the reaction at the anode is oxygen generation:

$2H_2O \longrightarrow O_2 + 4H^+ + 4e^-$ $E°ox = -1.23$ V

In the case where the anode is the metal to be plated, the plating bath is continuously replenished by the anode. In the case where the anode is an inert material, the ions of the metal to be plated should be periodically replenished in the plating bath.
The plating bath may contain other metal or non-metal compounds to improve the process, for example to improve conductivity or for other purposes. The metals to be plated are previ-

ously subjected to pretreatments to prepare the surfaces for plating.

There is a great variety of metals and alloys in electroplating, some using acid baths and some alkaline baths where the metals are found as cyanide complexes. Metals or alloys in electroplating include: brass (an alloy of Cu and Zn), cadmium, copper, nickel, zinc, gold, platinum, palladium, indium, rhodium, ruthenium, silver, tin, tin-lead alloys, tin-nickel, zinc-nickel, zinc-cobalt or zinc-iron. Below are discussed some metal plating along with the main techniques using ion exchange resins in recovering metals, plating baths, rinse waters or treating waste effluents.

Electroless plating

In electroless plating instead of an anode, the metal is supplied as the metal salt. Instead of a cathode, the substrate serves as cathode. The electrons to reduce the metal ion to the metal state are supplied by a reducing agent on the substate. Replenishment is achieved by adding metal salt.
A common reducing agent is hypophosphite, $H_2PO_2^-$. The reactions taking place are:

Oxidation (anodic):
$H_2PO_2^- + H_2O \longrightarrow H_2PO_3^- + 2H^+ + 2e^-$ $\quad E° = 0.50$ V
Reduction (cathodic):
$Ni^{2+} + 2e^- \longrightarrow Ni°$ $\quad E° = -0.25$ V
$2H^+ + 2e^- \longrightarrow H_2$ $\quad E° = 0$
$H_2PO_2^- + H^+ + e^- \longrightarrow P + OH^- + H_2O$ $\quad E° = 0.50$ V

The reaction of metal deposition must occur exclusively at the surface of the substrate. For this, various additives added, such as complzxing agents, contribute to avoid homogenious reactions. Subsequently, the formed metal film catalyzes further deposition so that the electroless process is also called autocatalytic process. Various additives are added to the plating bath in order to achieve the desired process conditions.
Additives added to the electroless plating baths are wetting agents, which are non-ionic surfactants, complexing agents, citric, maleic, oxalic acids or EDTA, which are added in order to maintain the metals concentration beyond their solubility, and stabilizers.

Chromium plating

The most common form of chromium plating is the hard chromium and the bright decorative chromium. Materials to be plated can be steel, brass, copper, zinc die-cast and aluminum. Pretreatments include degreasing, cleaning, various etching solutions, HCl pickling for steel, high alkaline pickling for aluminum. When plating hard chromium on steel, an underlying copper plating can be included in the process. Bright chromium layer is plated over a nickel underlayer. When plated on iron or steel or on zinc die-cast, an underlayer of copper allows nickel to adhere better.

Chromium plating is performed with chromates, CrO_4^{2-}, (Cr(VI) oxidation state) or with Cr^{3+}.

The cathode reactions that take place are:

$$Cr_2O_7^{2-} + 6\,e^- + 14\,H^+ \rightarrow 2Cr^{3+} + 7\,H_2O \qquad E°_{red} = 1.33\ V$$
$$2\,Cr^{3+} + 6\,e^- \rightarrow 2\,Cr° \qquad E°_{red} = -0.74\ V$$
$$\overline{Cr_2O_7^{2-} + 12\,e^- + 14\,H^+ \rightarrow 2Cr° + 7\,H_2O \qquad E°_{red} = 0.59\ V}$$

$$2\,H^+ + e^- \rightarrow H_2 \qquad E°_{red} = 0$$

The anode reactions are:
$$2\,H_2O \rightarrow O_2 + 4\,H^+ + 4\,e^- \qquad E°_{ox} = -1.23\ V$$
$$Cr^{3+} + 3\,H_2O \rightarrow CrO_3 + 6\,H^+ + 3e^- \qquad E°_{ox} = -1.33\ V$$

The chromate bath consists of a mixture of chromic acid and sulfuric acid at ratios $CrO_3:SO_4$ varying from 75:1 to 250:1. A fresh chromic acid bath contains 300-400 g/L chromic acid, 1-2% H_2SO_4, it can contain fluorides at 1-3% level and Cr^{3+} salts. During operation, this bath accumulates different metal ions depending on the surfaces that are being plated, such as Fe^{3+}, Ni^{2+}, Cu^{2+}, Zn^{2+} as well as Cr^{3+} originating from cathodic reduction of the chromate electrolyte. When the level of the metal impurities in the bath reaches a certain level and affects the quality of the plating, the bath should be either replaced or cleaned. A spent bath contains approximately 300 g H_2CrO_4/L and 10-25 g/L of metal impurities as mentioned above. The very high H^+ concentration of this bath makes the use of ion exchange, either a conventional SAC or a selective resin unfeasible. As the equation below indicates, at these H^+ concentrations the equilibrium with a SAC resin will be shifted to the left:

$$n\,R\text{-}SO_3^-H^+ + M^{n+} \leftrightarrows (R\text{-}SO_3^-)_n\,M^{n+} + n\,H^+$$

By diluting down the chromic acid solution, because we have here a mono-divalent equilibrium, the distribution of the polyvalent metal cations between resin and solution is in favor of the metals at low solution concentrations while it is in favor of the monovalent H^+ at high solution concentrations (pages 83-89). At a concentration of chromic acid of 100 g/L, the process using a SAC resin is feasible. An operating capacity of about 0.7 eq/L_R can be obtained at a specific flow rate of 3 BV/h and a regeneration level of the resin of 200 g HCl/L_R using a 10% HCl solution or 300 g/L_R H_2SO_4 using 15% H_2SO_4 solution. Under these operating conditions, a macroreticular type resin is advisable because these resins have a better oxidative as well as physical degradation resistance. After the SAC resin, the purified liquor is concentrated by evaporation and returns to the chromate bath. Figure 3.9 illustrates the chromate bath recovery process using a SAC resin.

A filter is placed before the resins to remove suspended matter. In general, two IER columns are placed in parallel, when one is on service the other is on regeneration or stand-by.

In recovering rinse waters at a chromium plating plant, it should be taken into account that rinse waters may come from other plating operations except chrome plating, such as Cu and Ni plating. These different platings may result into different rinse waters, acidic (chromium) or alkaline (Cu plating may be based on cyanide solutions). The treatment of the rinse waters can be done separately on acidic or alkaline rinse waters, or on mixed rinse waters making sure that the pH never becomes acidic during mixing causing the formation of HCN.

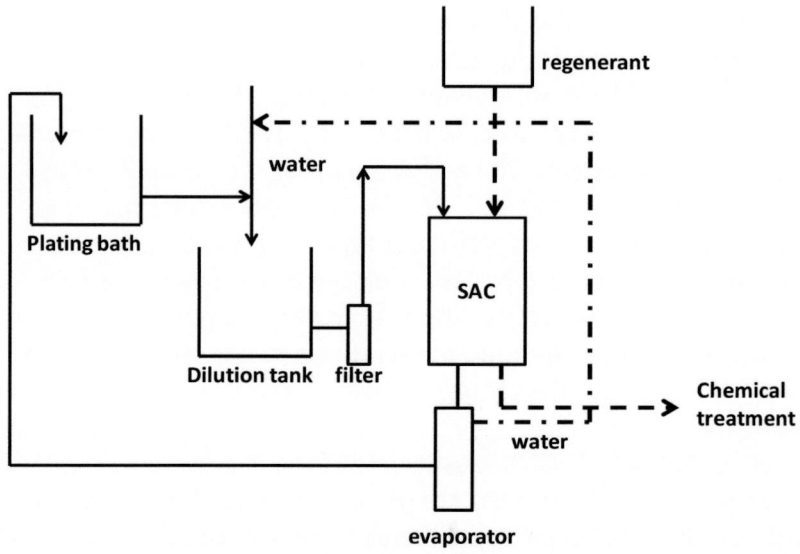

Figure 3.9 Chromate bath recovery

Figure 3.10 shows the flow diagram of the process of treating acidic rinse waters in chromium plating. The chromic acid rinse waters have a similar composition as the plating bath discussed above except that the concentration is much lower. Typically, these waters contain some 20-100 ppm chromic acid and 5-20 ppm of metals such as Fe^{3+}, Zn^{2+}, Cu^{2+}, Ni^{2+} and Cr^{3+}.

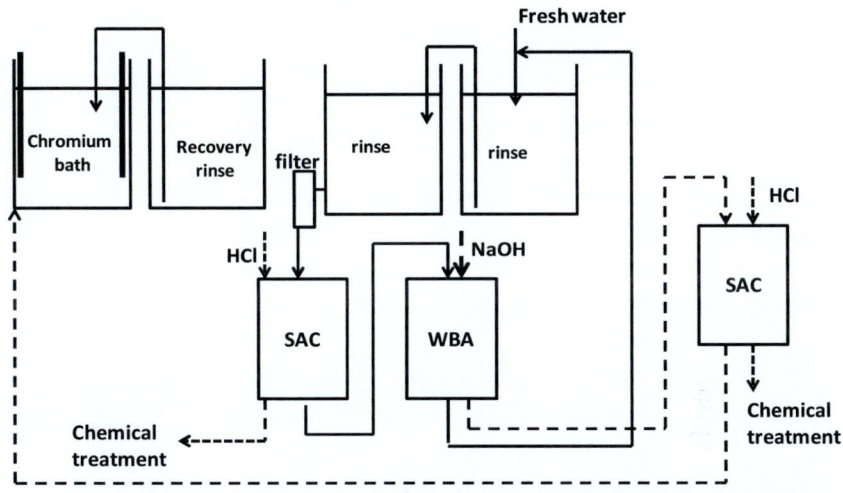

Figure 3.10 Acidic rinse waters recycling in chromate plating

The process consists of a SAC resin followed by a WBA resin followed by a SAC resin. As the chromium plating rinse waters pass through the first SAC resin, the metal impurities Fe^{3+}, Zn^{2+}, Cu^{2+}, Ni^{2+} and Cr^{3+} are fixed on the resin. The effluent contains essentially H_2CrO_4 with some H_2SO_4 which passes through the WBA resin where the acids are removed by the resin. The effluent from the WBA is then recycled to the rinse.

The flow rate in this process can be about 20 BV/h. The operating capacity obtained for the SAC resin is 0.8-1.0 eq/L_R with a regeneration level of 200 g HCl/L_R using 10% HCl or 300 g/L_R H_2SO_4 using a 15% H_2SO_4 solution. After regeneration, the SAC resin is washed with about 5 BV of water to displace the acid. The regeneration of the WBA is performed with 80-100 g NaOH/L_R using a 4% NaOH solution. The reason for the relatively high NaOH level for a WBA resin is that chromium may be found partially as dichromate, $Cr_2O_7^{2-}$, on the resin due to the acidic pH

$$2\ CrO_4^{2-} + 2\ H^+ \rightleftharpoons Cr_2O_7^{2-} + H_2O$$

and therefore, two NaOH equivalents may be consumed per one equivalent of dichromate:

$$(R-NH^+)_2\ Cr_2O_7^{2-} + 4\ NaOH \longrightarrow 2\ R + 2\ Na_2CrO_4 + 3\ H_2O$$

After regeneration the WBA resin is washed with 5 BV of water.

The spent regenerant from the WBA resin is essentially Na_2CrO_4 and unused NaOH. This liquor is then passed through the second SAC which is in the H^+ form where the Na^+ are fixed by the resin giving an effluent essentially chromic acid. The excess of NaOH is neutralized by the resin:

$$RSO_3^-\ H^+ + NaOH \longrightarrow RSO_3^-Na^+ + H_2O$$

This neutralization develops heat in the column and in presence of the chromic acid, it favors resin oxidation. For this reason, a macroreticular type resin is preferable here.

The chromic acid that comes out of the SAC has a concentration of about 50 g/L. It is subsequently evaporated and returns to the chromic acid bath.

Some chromates (Cr^{6+}) may be reduced on the WBA resin to Cr^{3+} which can then precipitate on the resin as hydroxides. For that reason, it is recommended to periodically perform an acid clean-up of both the SAC and WBA resins with 15% HCl.

As mentioned earlier, in a chromium plating shop different plating operations take place so that chromium, copper or nickel plating rinse waters co-exist in the same plant and copper or zinc cyanides are frequently found in such a plant. The flow diagram of alkaline or mixed rinse waters recycling is shown in figure 3.11.

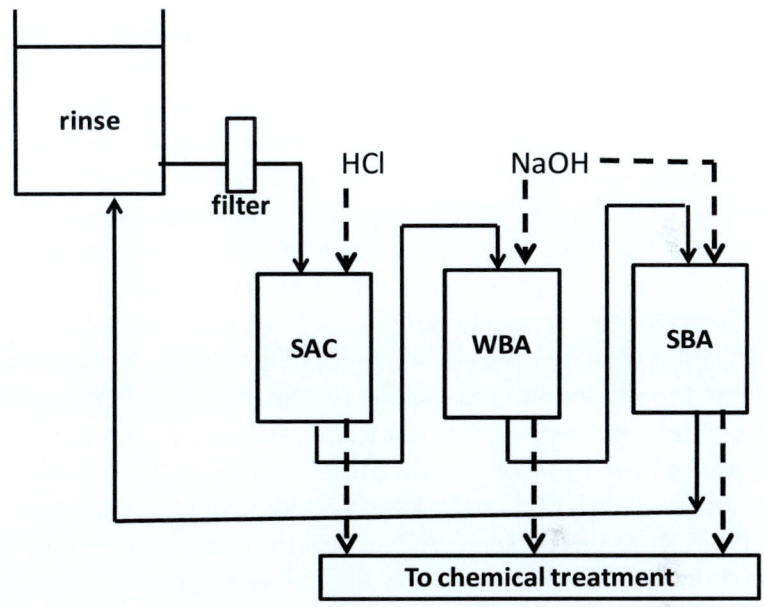

Figure 3.11 Alkaline or mixed rinse waters recycling

These mixed rinse waters contain metal impurities such as Cu, Zn, Ni, Fe and Cr as Cr^{3+} and as CrO_4^{2-}, they also contain CN^- either complexed with some of the metals or as free CN^-, plus some PO_4^{3-}, SO_4^{2-} and organic acids. The rinse waters pass first through the SAC resin where the free cations are fixed and where some of the cyanide complexes are dissociated to give metal cations (weak-acid dissociables (WAD) cyanide complexes which are relatively unstable complexes with metals such as

Cd, Zn, Ni and Cu to a certain extent), retained by the resin, and CN^-. The effluent of the SAC contain those metals that are strongly complexed with CN^- (strong-acid dissociables (SAD) complexes with metals such as Fe, Co, Au and Ag) strong acids such as H_2CrO_4, H_2SO_4 or H_3PO_4 and weak acids such as HCN. It should be pointed out that the presence of HCN does not constitute a danger for the health because HCN is very soluble in water and remains in solution. This effluent passes then through the WBA resin where the strong acids are fixed along with the CN^- complexes. The effluent of the WBA resin goes then to the SBA resin in the OH^- form where the weak acids such as HCN are fixed. The final effluents, usually having a conductivity of about 5 µS/cm are recycled to the rinse. An operating capacity of about 0.3 eq/L_R is obtained by the SBA resin. Regeneration is performed with 4% NaOH at a level of 100 g/L_R. It is recommended to regenerate independently the strong and the weak base anion exchange resins not to mix the CrO_4^{2-} stream from the WBA with the CN^- stream from the SBA resin. It is also recommended not to let the strong cyanide complexes break through from the WBA resin because they are very strongly fixed on the SBA resin and are difficult to regenerate, thus causing a fouling of the SBA resin.

Cr^{3+} plating has replaced to a certain extent the chromate plating due to the higher toxicity of Cr(VI) compared to Cr^{3+}. A typical trivalent chromium bath contains 5-15% chromium sulfate, 5-15% ammonium chloride, 5-15% sodium bromide, 5-15% boric acid, 1-5% ammonium formate and 0.5-1% sodium ethylhexylsulfate wetting agent. Impurities to remove in order to recover the bath are mainly Zn^{2+}, Ni^{2+}, Cu^{2+}, Fe^{3+}. Cr^{3+} plating baths cannot tolerate high concentrations of metal impurities. For example, in passivating of zinc plating with Cr^{3+}, impurities in the Cr^{3+} bath of Zn above 2000 mg/L and Fe above 100 mg/L cause problems in the coating.

The purification of Cr^{3+} baths and rinse waters is achieved with a selective resin capable to remove heavy metals such as Zn^{2+}, Cu^{2+}, Ni^{2+}, Fe^{2+}, Pb^{2+} but not Cr^{3+} cations at the pH and temperature of the plating bath. Such a resin is reported to be an IDA type (Reynolds, 1993). It is noted here nevertheless that IDA type resins have high affinity for Cr^{3+}, however, they need elevated temperatures, around 50°C, to fully elute this element. Purification of Cr^{3+} baths by removing Cu and Ni can be achieved with BPA resins (Dowex® M-4195). Elution is accomplished with 2N NH_4OH followed by water rinse and acidification with dilute H_2SO_4 to put the resin back to the SO_4^{2-} form.

Fe^{3+} was removed from spent Cr^{3+} baths using monophosphonic Purolite® S-957 resin. Better results were obtained at 55°C and slow flow rates. Regeneration of the resin was done with 6 BV of 30% HCl.

Printed Circuit Boards (PCB) rinse waters

Printed circuit boards (PCB) is a series of operations involving etching, plating and rinsing among many others, that generates various wastes. In these wastes are found heavy metals at high concentrations of which copper has a high recovery potential. Ion exchange plays an important role in recovering copper as well as removing metal impurities from copper solutions. Copper can be recovered from rinse waters, etching, plating or Chemical and Mechanical Polishing (CMP) of integrated circuit microchips using selective resins of iminodiacetic acid or picolylamine type or conventional ion exchange resins, depending on the characteristics of the various wastes.

In order to create copper paths on a copper-clad board, the desired pattern is drawn on the board or, for example, by ironing a printed pattern from a photographic paper, on the copper. The board is then etched by immersing it in a strong acid which dissolves all copper except in the protected areas, leaving the board with a copper pattern. As an etching solutions can be used $FeCl_3$, H_2SO_4/H_2O_2, H_2CrO_4/H_2SO_4, $CuCl_2$ or NH_4OH/NH_4Cl.

From ammonia-based etching solutions copper can be recovered by precipitation, SX or by electrolysis. At the pH of the spent solutions, 8.5-9.5, copper is found as $Cu(NH_3)_4^{2+}$. The spent solution, contains about 130-150 g/L of copper and 175-190 g/L chlorides. With the precipitation technology, the solution is first neutralized to a pH slightly acidic where copper precipitates out as $Cu(OH)_2$. After filtration to remove the $Cu(OH)_2$, the filtrate still contains about 3 g/L Cu which is recovered with an IDA type chelating resin.

Copper recovery from acidic rinse waters is achieved with a merry-go-round system, two columns on loading and one on regeneration with IDA type resins. In copper removal with IDA resins, the resin does not need to be in the Na^+ form. Because of the high affinity for Cu^{2+}, IDA resins can function in the H^+ form. When the second column starts leaking through, the first is practically exhausted and goes to regeneration. Regeneration is performed with sulfuric acid and the copper sulfate that is recovered is sent to an electrowinning system where copper is recovered as a metal.

As an illustration, with a feed solution of a pH=3, a copper concentration of 1 g/L and a specific flow rate of 20 BV/h, when the second column reaches a leakage of 10 ppm, the first column is loaded with 35-40 g Cu/L_R. By regenerating with 1.5 BV of 10% H_2SO_4, in the first 0.75 BV of the spent regenerant copper concentration is about 40-45 g Cu/L and is recovered while the second 0.75 BV contains only minor quantities of Cu and can be recycled.

Rinse waters from $CuCl_2$ etching containing a low concentration of copper, <50 mg/L and a pH not less than 2.5, then a strong acid cation exchanger can be envisaged (Hgiem et al, 2008). Regeneration was done with 12% HCl solution.
Chemical and Mechanical Polishing (CMP) of integrated circuit microchips consists in polishing and preparing the surface of the silicon wafer. It involves a polishing slurry which removes excess metal (copper usually) from the surface of the silicon wafer. The polishing slurry consists of an oxidant, such as H_2O_2, complexing agents, such as EDTA, ethylenediamine or citric acid, an abrasive, such as silica, and other ingredients. The copper introduced
in the polishing slurry must be removed in the waste effluent treatment of the semiconductor plant (Dungan and Han, 2001; Sassaman *et al*, 2004). After separating the slurry from the copper-bearing solution, by ultrafiltration for example, the solution passes through an ion exchanger of IDA type which removes copper. The system used is similar to that described in page 94 in Cu recovering from rinse waters. The resin can be regenerated with H_2SO_4 and the $CuSO_4$ eluate can be electrowon to recover metal Cu.
In the presence of EDTA or similar complexing agents, picolylamine resins are used as already mentioned above. Or

else, a PEI type resin (Maketon, 2007) which is polyethyleneimine immobilized on a solid support and which fixes copper even in the presence of complexing agents. The net charge of the complexed metal can be negative while the free metal ion has a positive sign. Resins such as PEI by their nature can fix both free Cu^{2+} as well as complexed Cu^{2+}, for example $[CuEDTA]^{2-}$.

Nickel recovery from plating rinse waters

Recovery of nickel dragged out of the nickel plating baths into the rinse waters can be recovered and recycled back to the plating bath using SAC exchange resins. There exist various techniques in nickel plating depending on the application. The plating baths contain usually nickel chloride and sulphate salts at a given pH, boric acid, plus other ingredients.
In order to recover the nickel salts from the rinse waters, the used waters pass through a three bed system consisting of a SAC, a WBA and a SBA exchange resin (Price and Novotny, 1979). The SAC is in the H^+ form where Ni^{2+} and other cations present are
fixed. The effluent of the SAC passes through the WBA resin where the strong acids are fixed and then passes through the SBA resin where the weak acids such as boric acid are fixed. The final effluent is deionized water and is recycled back to rinse.
The SAC resin is regenerated with H_2SO_4. The recovered spent regenerant is concentrated by evaporation and is sent for recovery of $NiSO_4$. The anion exchange resins are regenerated in series SBA ➔ WBA using 4% NaOH solution.

In another technique by Eco-Tec Inc., called NickelPur™, (Pajunen and Mangum) the rinse waters are treated with a SAC in the H^+ form where Ni^{2+} and other cations are fixed. The resin is then regenerated with H_2SO_4. Since SAC have higher affinity for divalent cations over monovalent, in order to favour the removal of Ni^{2+} over monovalent cations like Na^+, the cycle can be extended so that Na^+ leak through the resin displaced by Ni^{2+} ions. The regeneration eluate is collected and passed through a de-acidification unit containing a SBA resin, where the nickel salt is separated from the free acid. This de-acidification unit is based on the acid retardation principle and has already been described earlier (page 70).

Copper recovery from plating rinse waters

Copper recovery from acidic rinse waters can be achieved in a similar way (SAC ➔ WBA). An efficient configuration consists of a filter to remove suspended matter, activated carbon to remove organic additives followed by the SAC and weak base anion exchangers. The resin is regenerated with H_2SO_4 and copper is recovered by EW. A process to recover copper with a SAC resin that uses electrochemical regeneration (Jones *et al*, 2006) was reported. The cation exchange resin was regenerated by applying an electric field through the resin and concentrate copper in the concentrate which is then treated with conventional means. Copper content was reduced from 40 ppm in the feed to 1.8 ppm in the treated solution.

Cadmium electroplating

Cadmium is electroplated more frequently from cadmium cyanide baths even though there exist non-cyanide processes. Rinse waters from the second rinse tank contain about 10 mg/L cadmium and 35 mg/L cyanides which can be removed using a SBA exchange resin. These waters are first filtered to remove any suspended matter and then go to the IX column. When the resin is exhausted it is regenerated with 15-20% NaOH and the spent regenerant goes to EW to recover cadmium. The treated water returns to rinse.

Another way that has been reported is to use an IDA type resin in the Na^+ form which removes cadmium from cyanide solutions (Koff and Zarate, 1997). Cadmium forms a "weak" complex with cyanide and apparently IDA resins fix Cd^{2+} from cyanide solutions. The resin is regenerated with acid followed by Na^+ conversion. This is very interesting because the stability constant of $[Cd(CN)_4]^{2-}$ is much higher than the stability constant of Cd^{2+}-IDA complex at pH>7.

Galvanization

Galvanization is the deposition of a coating of zinc on steel or iron in order to protect them from rust. The zinc coating is applied either with the hot-dip technology where the steel parts are immersed in a bath of molten zinc, or with electrodeposition. In the electrodeposition technology, the plating bath is a zinc solution (zinc chloride or zinc sulphate in acidic baths), either with pure zinc or with alloys like Zn/Fe, and a consumable zinc anode. Zinc electroplating provides corrosion resistance to the steel by acting as barrier and sacrificial coating. Pretreatments include degreasing and pickling. Various post-treatments can be applied to improve anti-corrosion and paint properties such as chromate (Cr^{3+}) or phosphate conversion coatings (page 103).

In electrogalvanization, the Zn in the anode dissolves continuously into the bath while Zn from the bath is deposited on the steel surface. However, the rate of dissolution is higher than the rate of deposition so that the Zn concentration in the bath increases. In order to avoid that Zn reaches a concentration such that the quality of the coated steel will be affected, an ion exchange system can be used in the case where the bath contains zinc chloride. The system consists of a strong base anion exchange resin in the Cl^- form which removes Zn in the form of $ZnCl_4^{2-}$. If iron is present in the case of Zn/Fe alloy, it is not removed when it is found in the Fe^{2+} state since Fe^{2+} does not form complexes with Cl^-. The resin volume used is such as to remove the excess of Zn only.

As galvanization goes on, the plating bath becomes contaminated with Fe^{2+} from the steel or iron parts. The plating bath can be purified and recycled using a Fe^{3+} selective, iminodiacetic type, ion exchange resin. The high metal selectivity of this type of resins for Fe^{3+} is used to purify galvanizing solutions from Fe^{3+} impurities.

A typical solution contains about 90 g Zn^{2+}/L Fe^{2+} 0.7 g/L, Fe^{3+} 1 g/L, Ni^{2+} 0.2 g/L, Na_2SO_4 55 g/L and a pH of 1.7. Fe^{2+} is oxidized to Fe^{3+} using H_2O_2 and then the solution is passed through IDA type resin in the H^+ form at a flow rate of 6 BV/h. It should be made sure that there is no excess H_2O_2 in the solution before passing through the IER in order to avoid resin oxidation. The average iron concentration in the treated solution is decreased by about one half while the resin, even at these highly unfavourable conditions of high ionic background and low pH, removes about

13 g Fe^{3+}/L_R. Regeneration was performed using 2 BV of 15% H_2SO_4.

Zn/Ni electrolytic solutions can also be treated, where Zn^{2+} and Ni^{2+} concentrations are around 50 g/L each, Na_2SO_4 around 80 g/L, total iron around 1 g/L and a pH at 1.5-1.9. Fe^{3+} is removed with an IDA type resin, as above. Operating capacities of 8-10 g Fe/L_R can be obtained under similar conditions as above. Figure 3.12 illustrates the Fe3+ removal from a Zn/Ni electrolytic solution using an IDA type resin.

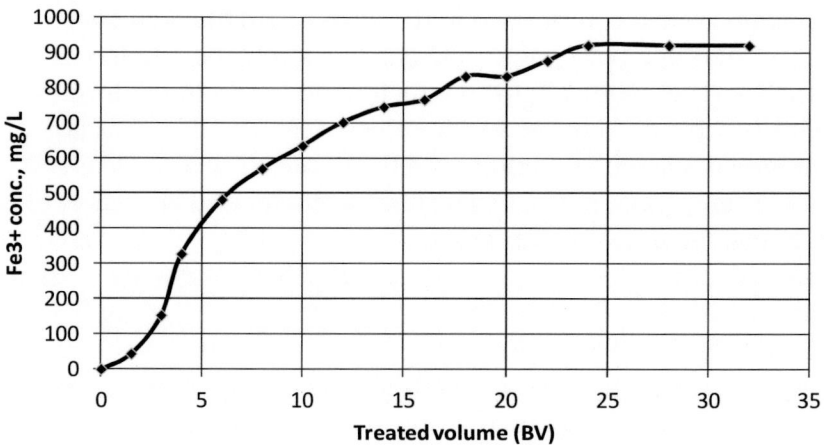

Fig. 3.12 Fe^{3+} removal from Zn/Ni electrolytic solution. Feed: Fe^{3+} 1 g/L, Zn^{2+} 48 g/L, Ni^{2+} 57 g/L, Na_2SO_4 70 g/L, pH=1.5

Rinse waters from galvanizing steel or iron containing about 100 ppm Zn^{2+} at a pH of about 4-6 have been treated with a gel type SAC resin (Maranon et al, 2005). At this concentration of Cl^-, Zn is found as cation Zn^{2+} and is removed by the SAC resin in the H^+ form.

Electroless copper and nickel recovery

The most frequent electroless plating is with nickel. Nickel sulfate is used as the metal salt. Cu or Ni from electroless Cu or Ni plating where complexing agents such as alkanolamines, ammonia or carboxylic acids are used in the bath, IDA type resins have been found to remove Cu or Ni from baths or rinse waters and leave the complexing agents in solution (Courduvelis and Gallager, 1981). The effluents of the chelating resin go to the waste treatment while the metals can be recovered by elution with H_2SO_4. The process does not work when EDTA or nitriltriacetate (NTA) are used as chelating agents because Cu forms a stronger complex with these two chelators than with the iminodiacetic groups of the resin.
In a similar manner, Ni^{2+} can be removed and recovered from rinse waters with IDA type resins, in a three columns merry-go-round system. In this case, the resin should be in the monosodium form. When the first column is exhausted, most of the Na^+ ions have been displaced by the Ni^{2+} ions.

The conventional way to remove metals from wastes is to raise the pH so that metals precipitate out as hydroxides. If chelating agents such as EDTA are present, they keep the metals in solution and prevent from precipitation. One way to solve this problem is to lower the pH and add an iron salt. The complex metal-EDTA is weakened and Cu is replaced by Fe. Cu is then precipitated by raising the pH. With this technique however the precipitated Cu is contaminated with Fe and is therefore of no value. When complexing agents like EDTA are used in the bath, IDA resins cannot function properly as mentioned above. In that case, a picolylamine type resin (BPA or HPPA) can be used (Brown and Dejak, 1987). The BPA resin has a strong affinity for copper at low pH in presence of EDTA chelator and it can selectively remove copper in presence Fe^{2+} or Fe^{3+} which remain in solution, probably complexed with EDTA. The BPA resin is regenerated with high concentration (10N) H_2SO_4 while HPPA is regenerated with 2N H_2SO_4.

Another way to remove copper in presence of EDTA as chelator is to use a polyethyleneimine (PEI) type resin (Matejka and Eliasek, 1987). According to these authors, the PEI resin is capable to fix Cu or Ni as free cations with the imine functional groups while EDTA is held on the resin by electrostatic attraction with the positively charged metal cation. The regeneration is done first with a salt to remove the electrostatically bound EDTA, followed by an acid elution of the metals and finally with a caustic wash to convert the resin into the free base form.

Rinse waters containing copper-EDTA complex has been treated with a SBA exchanger to remove both free EDTA and the Cu-EDTA complex (Sricharoenchaikit, 1989). Regeneration of the resin was done with a 10% NaCl solution. Both, Cu and EDTA were recovered and recycled.

Gold plating

Gold plating techniques include alkaline gold cyanide, neutral gold cyanide, acid gold cyanide and non-cyanide plating (sulphite or chloride based). In alkaline gold cyanide baths there is also free cyanides present so that gold metal can be used as a consumable anode. Gold is found in the bath as cyanide complex $Au(CN)_2^-$. The reactions that take place are:

Cathode: $\quad Au(CN)_2^- + e^- \longrightarrow Au + 2\ CN^-$

Anode :	$Au + 2\ CN^- \longrightarrow Au(CN)_2^- + e^-$

If non-soluble anodes are used, like stainless steel, the reactions are:

Cathode:	$4\ Au(CN)_2^- + 4\ e^- \longrightarrow 4\ Au + 8\ CN^-$
Anode :	$2\ H_2O \longrightarrow 4H^+ + O_2 + 4\ e^-$

Overall: $4\ Au(CN)_2^- + 2\ H_2O \longrightarrow 4\ Au + 8\ CN^- + O_2 + 4\ H^+$

The rinse waters contain low concentrations of gold cyanides, too low to recover by electrochemical methods. The use of ion exchange resins to recover gold in plating operations is accomplished with anion exchange resins. The technique of using anion exchange resins to recover gold and silver cyanides from plating baths or rinse waters is known since 1948 (Byler and Dunn, 1953) where a polyamine phenol-formaldehyde resin was employed (Duolite® A-3). Loading was performed at a pH of 8.5-10.5 and regeneration was done with 1% NaOH. Today, gel type SBA resins are employed. Two columns in series are used, the second being a polisher. When the first column breaks through, the resin is removed out and a new resin is placed in the polishing position. Since regeneration of the resin is very difficult, the common practice to recover the gold from the resin is to dry the resin and incinerating it. The value of the gold fixed by the resin (the resin operating capacity can be around 100 g/L_R, depending however on the composition of the solution) by far justifies the sacrifice of the resin. If large quantities of resin are involved regeneration techniques exist, as for example (Lawn, 1983) the elution with an eluent consisting of an organic solvent, water and potassium thiocynate.

Acidic cyanide baths use $KAu(CN)_2$ as the source of gold and use a citrate buffer to obtain a pH of 4. From acidic solutions where citric acid is used to keep gold in solution, thiouronium resins have been suggested (ResinTech, 2008) due to the fact that conventional SBA resins have low capacity under these conditions.

Silver recovery

In plating waste streams, silver is generally found as silver cyanide complex, $Ag(CN)_2^-$. Using a gel type SBA resin it was possible to remove $Ag(CN)_2^-$ from solution with a capacity of 1.1 eqL_R (Burstal et al, 1953). Regeneration was performed with 2N potassium thiocyanate. Silver is recovered from the spent regenerant by electrolysis. Another way, depending on the price of silver, is to recover silver by incinerating the resin.

Platinum group metals

Platinum group metals (PGM) include Pt, Pd, designated as primary PGM, Rh, Ir, Ru and Os, designated as secondary PGM. In all HCl concentrations, Pt and Pd are found as $PtCl_6^{2-}$, $PtCl_4^{2-}$ and $PdCl_4^{2-}$ whereas secondary PGM are found as mixtures of chloro- and aquochloro-complexes ($RhCl_6^{3-} \leftrightarrows RhCl_5(H_2O)^{2-}$) (Green et al, 2004). The aquochloro- complexes are favored by HCl concentrations below 6N (Kononova et al, 2010). Base metals such as Cu^{2+} and Fe^{3+} form also chlocomplexes in equilibrium with neutral and cationic form complexes ($FeCl_4^- \leftrightarrows FeCl_3(H_2O)_3 \leftrightarrows FeCl_2(H_2O)_5^+$) which are favored by low acid concentrations. The affinity of a SBA resin for the above chlorocomplexes is therefore expected to be:

    Primary PGM > secondary PGM > base metals

especially at low acid concentrations. It was in fact found that anion exchange resins showed reduced affinity for base metals at HCl concentrations below 3N allowing the removal of PGM in presence of base metals. The resin Reillex® HP 425, a pyridine resin, showed the best loadings, with thiourea resins next (Green *and al,* 2004).

Pd(II) is used in the PCB industry in acidic solutions where it is found as an anionic complex, $PdCl_4^{2-}$. Platinum is found in waste streams of plating shops as the anionic complex $PtCl_6^{2-}$. A SBA resin is used for its high selectivity for these complexes for both cases. Due to the high affinity of the resin, regeneration is difficult and resin incineration is practiced, as is the case with gold recovery.

Rhodium from acidic media is characterized by the kinetic inertness of its complexes (Kononova *et al*, 2010). Therefore its recovery, especially in presence of other precious or non-ferrous metals is incomplete. For that reason, recovery of rhodium is based on both, SBA resins where the Rh complexes are attracted by the resin by electrostatic forces and on selective resins bearing N, O, or S containing ligands where the Rh forms coordination bonds. For example, the simultaneous recovery of Rh and Pt from acidic solutions was studied (Kononova *et al*, 2011) using resins with N-bearing functional groups, including SBA, WBA, pyridine and polyamine resins. The parameter investigated was the acidity of the solutions. It was found that the best resin for Rh recovery in presence of Pt was the polyamine resin. The WBA resins showed better recoveries for both Rh and Pt at higher pH (low acidity) due to the deprotonation of the amine group and the formation of covalent bonds with the PGM. The SBA showed better performance at low acidity due probably to

the lower Cl⁻ concentration and therefore the PGM anionic complexes has less competition from the Cl⁻ anions.
Recovery of PGM with resins containing S has been applied based on the fact that PGM form stable complexes with S-containing ligands (Hubicki *et al*, 2008).

Organics removal from rinse waters

In a number of baths, organic compounds are added as wetting agents or brighteners. These compounds include ionic or non-ionic surfactants that risk to foul the resins. For example, cationic surfactants may foul cation exchangers and anionic surfactants may foul anion exchangers. These compounds should therefore be removed before the ion exchange resins.
There exist various techniques to remove such compounds. One technique is adsorption, either with activated carbon or with synthetic adsorbents. The ability of synthetic adsorbents to remove these compounds from baths or rinse waters depends on the chemical composition of these molecules and it is a common practice to test an adsorbent beforehand. Frequently, adsorbents having small size pores have been found suitable.

Wastes from metal plating industries

Effluent solutions originate from baths, recovery rinse or rinse waters of the various operations on metals discussed in this chapter: plating, anodizing, conversion coating, etching or PCB effluents. These effluents go finally to the chemical treatment plant from where they are discharged to the environment. Ion exchange is used at this point to render the effluent conformed to the existing regulations for the effluent waters quality.

First, any hexavalent chromium should be reduced to trivalent and any cyanides must be destroyed (oxidized) before any further treatment. Figure 3.13 illustrates a chemical treatment of galvanic rinse waters.

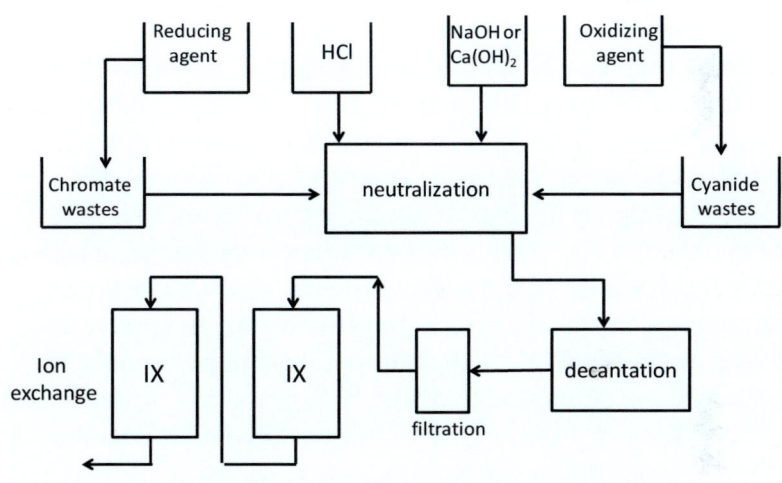

Figure 3.13 Waste waters purification

The chemical treatment consists of reducing Cr^{6+} (CrO_4^{2-}) to Cr^{3+} while the cyanides, CN^-, are oxidized to CO_2 and N_2. The wastes are then collected and neutralized in a neutralization tank where the metallic precipitates are filtered, pressed and disposed as solid wastes. Since the final effluent may still contain some low levels of heavy metals, a polishing unit containing a selective IER is frequently included. The filtrate after the decantation step contains about 5 to 20 ppm of heavy metals, in addition to higher concentrations of Na^+ and Ca^{2+} salts, in the order of several grams per litre, coming from the neutralising chemicals. metals frequently found here are Cd, Cr, Cu, Pb, Ni, Zn.

Because of the high concentration of Na^+ and Ca^{2+}, the removal of the metal impurities down to low levels (below 1 ppm) from the effluent becomes difficult with a conventional SAC resin. In that case, a selective resin of the IDA type is usually employed in a two-columns-in-series, lead-lag configuration, which can bring the total concentration of metals below 0.1 ppm.

The loading step can be done at a flow rate of 20-30 BV/h while regeneration can be performed at 120 g HCl/L_R regeneration level using 5 to 10% HCl, or the equivalent quantity of H_2SO_4. After the acid elution, the resin is converted back to the Na^+ form. Depending on the NaOH quantity per liter of resin, the conversion to the Na^+ form may be total or only partial. The degree of conversion of the resin to the Na^+ form as well as the influent composition will affect the effluent pH. In applications involving metal removal, with solutions containing mainly Na^+ salts, the resin is converted to about 50% in the Na^+ form. In that case, the effluent pH initially is slightly acidic and it becomes progressively neutral as the resin is loaded with the influent Na^+ and Ca^{2+} ions. With solutions containing mainly Ca^{2+} salts, the resin is converted more fully (70% or more) in the Na^+ form. In that case, the effluent pH initially is alkaline and becomes progressively neutral as the resin is loaded with the influent hardness ions. In order to distribute more uniformly these 50 or 70% of the sites in the Na^+ form, the resin should be conditioned under agitation.

If the metals are complexed with EDTA for example, they do not precipitate as easily as uncomplexed metals. In that case IDA type resins are not adequate to remove them and other selective resins should be used such as bis-picolylamine type (BPA or HPPA) or polyethyleneimine (PEI) type resin.

Summary of ion exchange systems used in metal finishing

Summarizing what has been discussed in this section, we have the following ion exchange systems that can be used in the metal processing industries.

I. A WAC resin to remove metal cations from water at neutral pH and low salt background.

II. A single SAC resin to remove cation metal impurities from acidic baths. For example, chromate bath recycling, or nickel removal from nickel electroplating baths, or purification of H_2SO_4 pickling liquor.

III. A single SBA resin for the removal of metals as anionic complexes. For example, gold recovery from alkaline cyanide liquors, $ZnCl_4^{2-}$ removal from HCl pickling liquors, or Fe^{3+} and Zn^{2+} removal in the purification of concentrated HCl.

IV. A SBA resin to separate an acid from its salt (acid retardation), see for example aluminum anodizing bath recovery.

V. Conventional water demineralization system:

SAC → WBA → SBA or SAC → SBA or SAC → WBA

These systems can be used for rinse waters recycling, as for example alkaline or mixed rinse waters or rinse waters recycling in nickel electroplating.

VI. Selective resins:
- IDA for selective removal of most transition metals, tolerate high Ca^{2+} background
- AMP: for selective removal of most transition metals except Hg^{2+} and Fe^{3+} which foul this type of resins. It has higher selectivity for Ca^{2+} than IDA resins.
- BPA for Cu^{2+}, Ni^{2+}, Zn^{2+} in presence of Cr^{3+} at low pH and for metals complexed with strong complexing agents (EDTA)
- Diphonix resins, high selectivity for Fe^{3+} with respect to other metals such as Cu^{2+}, Ni^{2+}, Co^{2+}.
- Thiol, thiourea or thiouronium resins for Hg, Au, Ag, Pt-group metals
- D2EHPA impregnated resins, selective for Zn^{2+}, Fe^{3+}, UO_2^{2+}, rare earths.

Examples:
- Cu^{2+} recovery from PCB rinse waters using IDA resins
- Removal of Fe^{3+}, Ni^{2+}, Zn^{2+} from Cr^{3+} plating baths with BPA resins or Diphonix resins fot Fe^{3+}
- removal of Zn^{2+} and Fe^{3+} from Cr^{3+} passivation baths with D2EHPA impregnated resins.
- removal of Fe^{3+} from galvanizing baths with IDA resins
- Removal of metals from high salt background waste waters with IDA resins

- Removal of Cu and Ni complexed with EDTA from electroless plating baths with BPA resins.
- gold recovery from acidic solutions with thiouronium or thiol resins

Cu and V recovery from adipic acid

Adipic acid, $HOOC\text{-}(CH_2)_4\text{-}COOH$, is an important dicarboxylic acid whose main use is the manufacturing of nylon-6,6 by a polycondensation reaction with hexamethylene diamine. Adipic acid is produced by oxidation of the KA oil with HNO_3. The KA oil is a mixture of cyclohexanol and cyclohexanone (KA stands for *K*etone-*A*lcohol). Upon oxidation with HNO_3, nitrous oxide, N_2O and NO_x are formed which are regulated. Copper and vanadium salts are used as catalysts. Adipic acid is crystallized and separated from the mother liquor which contains the copper (8000-11000 ppm) and vanadium (600-2500 ppm) ions along with nitric acid (less than 20%) and dicarboxylic acids by-products. Copper and vanadium are removed from the mother liquor by using a SAC resin (Monsanto Chemicals, 1964) especially made to resist oxidation in this highly oxidative environment. Copper and vanadium are then recovered by eluting with 25-30% nitric acid at a level of 600 g HNO_3/L_R. The eluate is recycled back to the oxidation reaction.

With the above composition of the feed solution to the resin, the operating capacity is about 3 BV. During the recovery of the metals, it has been observed that although Cu^{2+} is removed easily with the SAC resin, vanadium is more difficult to remove giving frequently high leakage and low capacity. This has been attributed to the possibility that vanadium, being in the V^{5+} state, is found in the solution as both cation (VO_2^+) and anion (vanadates, VO_4^{3}). In a suggested process (Hsu and Laird, 1984) SO_2 is introduced into the mother liquor which results in a better removal of vanadium, probably because V^{5+} is reduced to V^{4+} which is better fixed on the SAC resin as VO^{2+}.

Zn from viscose rayon spinning effluents

In spite of production decrease of viscose fibers in Europe since the end of last century, the rest of the world production has been increasing since the beginning of this century especially in Asia (Bywater, 2011).

The viscose solution is extruded through numerous spinnerets into a spinning bath which contains 8-10% H_2SO_4, 17% Na_2SO_4, 0.5-1% $ZnSO_4$ and some other ingredients. The spun yarn is then washed thoroughly with water.

The process waste waters contain mainly H_2SO_4, NaOH, sulfides, zinc and organics, essentially hemicelluloses .The acidic spent water effluents contain H_2SO_4, Na_2SO_4 and $ZnSO_4$ but at lower concentrations than in the bath. Typical acidic effluents have the following composition:

Acidic wastes
Zn^{2+} 180-315 mg/L or about 5-10 meq/L
SO_4^{2-} : 8400-32200 mg/L or 175-670 meq/L
Acidity: 2200-7700 mg CaCO3/L or 50-150 meq/L
TDS: 8900-32000 mg/L
pH=1-2

Combined effluents have the following composition:

Zn^{2+}: 2-18 mg/L
SO_4^{2-}: 700-6700 mg/L
Cl^-: 37-200 mg/L
Acidity: 160-1000 mg $CaCO_3$/L
TDS: 1100-1200 mg/L
pH=2.8-7.3

depending on the effluents composition, a SAC resin can be used to remove Zn^{2+}. Regeneration can be done using spinning liquor. If HCl is used as regenerant, then Zn^{2+} can be recovered from the spent regenerant as a $ZnCl_4^{2-}$ complex with a SBA resin.
Another option to remove Zn^{2+} from the effluents is by using an IDA type chelating resin, after adjusting the pH, if necessary, to about 3-4.

Copper and ammonium from cuprammonium rayon spinning effluents

Cellulose is dissolved in a solution containing caustic, ammonia and copper sulfate. The cellulose is actually dissolved in $[Cu(NH_3)_4](OH)_2$ and then regenerated as rayon by extrusion through a spinneret in sulfuric acid. The spinning effluents contain large quantities of ammonia, copper and sulfuric acid which can be removed with ion exchange resins.
The ion exchange technique depends on the composition of the effluents to treat. If the pH of the effluents is acidic, that is the composition is essentially H_2SO_4, $CuSO_4$ and $(NH_4)_2SO_4$, then a

SAC resin in the NH_4^+ form is used. The resin removes copper and it is regenerated with $(NH_4)_2SO_4$:
Loading:
$$2\ R\text{-}SO_3NH_4 + Cu^{2+} \longrightarrow (R\text{-}SO_3)_2Cu + 2\ NH_4^+$$
Regeneration:
$$(R\text{-}SO_3)_2Cu + (NH_4)_2SO_4 \longrightarrow 2\ R\text{-}SO_3NH_4 + CuSO_4$$

If the pH is neutral, a WAC resin in NH_4^+ form is used to remove Cu^{2+}. The resin is then regenerated with H_2SO_4 and put back to the NH_4^+ form using spent effluents containing NH_4OH.
Loading:
$$2R\text{-}COO^-NH_4^+ + CuSO_4 \longrightarrow (R\text{-}COO^-)_2Cu^{2+} + (NH_4)_2SO_4$$
Regeneration:
$$(R\text{-}COO^-)_2Cu^{2+} + H_2SO_4 \longrightarrow 2R\text{-}COO^-H^+ + CuSO_4$$
$$R\text{-}COO^-H^+ + NH_4OH \longrightarrow 2R\text{-}COO^-NH_4^+ + H_2O$$

With basic cupric effluents, a WAC resin in the H^+ form is used. During loading, both NH_4^+ and Cu^{2+} are fixed on the resin. Regeneration is done in two steps. In the first step, dilute (spent) H_2SO_4 is used to elute NH_4^+. In the second, more concentrated H_2SO_4 is used to elute Cu^{2+}.
Loading:
$$3\ R\text{-}COO^-H^+ + 3\ NH_4OH + CuSO_4 \longrightarrow (R\text{-}COO^-)_2Cu^{2+} +$$
$$R\text{-}COO^-NH_4^+\ (NH_4)_2SO_4 + 3H_2O$$
Regeneration:
$$2R\text{-}COO^-NH_4^+ + H_2SO_4\ (\text{dilute}) \longrightarrow 2R\text{-}COO^-H^+ + (NH_4)_2SO_4$$
$$(R\text{-}COO^-)_2Cu^{2+} + H_2SO_4 \longrightarrow 2\ R\text{-}COO^-H^+ + CuSO_4$$

4. Hydrometallurgical applications

Ion exchange resins can be used in hydrometallurgy to recover metals from leach liquors, to remove (metal) impurities from leach liquors and electrolytes and to recover metals from low grade ores, secondary streams, wastes or solvent extraction (SX) raffinates. The development of special resins allowed the purification of advance electrolytes and the resin-in-pulp (RIP) technology can be used to achieve in a cost effective way the recovery of base metals from pulps that have poor filterability or poor settling characteristics.

Today, the main applications of ion exchange resins in hydrometallurgy concern the recovery of uranium and gold from leach liquors. However, reviewing IER developments shows that more hydrometallurgical processes can be developed with ion exchange for metals such as PGM, base non-ferrous metals (copper, zinc, cobalt, nickel), molybdenum, rhenium, tungsten, vanadium and rare earths.

Uranium recovery with IER has already been the subject of a book of this series (Zaganiaris, 2009) and will not be discussed further. It will only be mentioned here that recently, AMP type resins have been introduced to recover uranium from acid leach liquors containing high Cl^- concentrations (Rezkallah, 2012;

Carr *et al*, 2012), thus avoiding the use of SX in these cases. According to this approach, pregnant leach solutions (PLS) containing up to 20 g/L of chlorides can successfully be treated with AMP type resins giving operating capacities of 20 g U/L_R at flow rates of 2.5 BV/h. Elution with NH_4OH and Na_2CO_3 is applied.

In the rest of this chapter it is reviewed existing or potential processes in hydrometallurgy for various metals where ion exchange can be part of the process.

Gold recovery

Leaching

The dissolution of gold in aqueous solution is achieved by oxidizing first Au° to an ionic species followed by complexation to stabilize it in solution. Gold forms stable complexes with "soft" electron donor ligands, CN^-, thiourea, thiocyanate, thiosulfate, which prefer low valency gold Au(I) and with "hard" which prefer high valency gold, Au(III), $AuCl_4^-$.

The choice of the leaching chemistry, cyanide, thiourea, thiosulfate or halide, depends on the ore type and safety considerations.

Gold is leached most frequently by alkaline cyanides with dissolved oxygen as the oxidizing medium:

$$4\ Au° + 8CN^- + 2\ H_2O + O_2 \longrightarrow 4\ Au(CN)_2^- + 4OH^-$$

Leaching can be done by a solution of NaCN and $Ca(OH)_2$ in vessels where milled ore is leached under agitation. The

Ca(OH)$_2$ prevents the decomposition of NaCN to HCN and reduces the amount of NaCN to leach gold. When gold price is depressed, low cost heap leaching is also practiced, giving low grade clear gold solutions.

Refractory ores, that is, ores resistant to cyanide leaching, usually contain sulfide minerals which trap gold particles and make them difficult to leach. These ores require pretreatment before cyanide leaching, such as concentration by sulfide flotation, roasting to oxidize sulfide minerals and others. Instead of roasting, because of SO$_2$ gas emissions and the necessary control, other techniques have been developed like pressure oxidation and bacterial leaching.

Cyanide leaching is the standard of the industry. Because however of certain drawbacks of the cyanide leaching, other leaching techniques have been developed including thiosulfate, thiourea and halides (Cl-, Br-, I-) leaching. Alkaline sulfide leaching has also been studied (Jeffrey and Anderson, 2003). Drawbacks of cyanide leaching are: non-adapted to certain ores, regeneration of CN$^-$ solutions (recycling CN$^-$, produce cleaner tailings) and toxicity of cyanide compounds.

Thiosulfate leaching is suitable for copper containing ores where copper consumes significant quantities of cyanides in cyanide leaching. Ammonium thiosulfate leaching (Araki *et al*, 2004) is a promising route because cupric ions present in the leach liquor from the ore is used as gold oxidant.:

$$Au^\circ + 4\ S_2O_3^{2-} + Cu(NH_3)_4^{2+} \longrightarrow Au(S_2O_3)_2^{3-} + Cu(S_2O_3)_2^{3-} + 4\ NH_3$$

Copper (I) is converted back to copper (II) by oxygen:

$$4\ Cu(S_2O_3)_2^{3-} + O_2 + H_2O + 16\ NH_3 \longrightarrow 4\ Cu(NH_3)_4^{2+} + 8\ S_2O_3^{2-} + 4\ OH^-$$

Thiourea leaching is done in acidic pH and Fe^{3+} is the oxidant for $Au°$. Gold is dissolved as a cationic complex.

$$2\ S{=}C(NH_2)_2 + 2\ Fe^{3+} \xrightarrow{H+} [(H_2N)_2{-}C{-}S{-}S{-}C{-}(NH_2)_2]^{2+} + 2Fe^{2+}$$

$$[(H_2N)_2{-}C{-}S{-}S{-}C{-}(NH_2)_2]^{2+} + 2\ S{=}C(NH_2)_2 + 2\ Au° \longrightarrow$$

$$2\ [Au(S{=}C(NH_2)_2)_2]^+$$

Leaching with halides, especially Cl^- and Br^-, in acidic conditions is practiced with refractory ores where conventional cyanide leaching is not efficient, or with ores containing carbonaceous materials that act as "preg-robbers". The reactions of halide leaching are as follows:

$$Cl_2 + H_2O \Leftrightarrow HCl + HOCl$$
$$3\ HCl + 2\ Au° + 3\ HOCl \leftrightarrows 2\ AuCl_3 + 3\ H_2O$$
$$AuCl_3 + Cl^- \Leftrightarrow AuCl_4^-$$

Thiocyanate leaching is done at 0.01-0.05 M SCN^- at potential 0.4-0.45 V and at pH=1-3 in the presence of ferric ions (2-5 g/L):

$$Au° + 4\ SCN^- + 3Fe^{3+} \leftrightarrows Au(SCN)_4^- + 3\ Fe^{2+}$$

Both, $[Au(I)(SCN)_2]^-$ and $[Au(III)(SCN)_4]^-$ can be formed depending on the potential.

Gold recovery from pregnant leach solution (PSL)

Anion exchange resins, weak or strong base, are used to fix the aurocyanide complex. The mechanism for loading a SBA resin is:

$$R\text{-}N^+R_3\ X^- + [Au(CN)_2]^- \leftrightarrows R\text{-}N^+R_3[Au(CN)_2]^- + X^-$$

The mechanism for loading a WBA resin is first protonate the resin and then fix the aurocyanide complex:

$$R\text{-}NR_2 + H^+X^- \longrightarrow R\text{-}NR_2H^+X^-$$
$$R\text{-}NH+X- + [Au(CN)_2]^- \leftrightarrows R\text{-}NH^+[Au(CN)_2]^- + X^-$$

However, the WBA resins can be protonated at a pH lower than their pKa value. Typical styrenic WBA resins have a pKa of about 8-9 while the pH of a cyanide PLS has a pH of about 10.5. Consequently, in order to fix the aurocyanide complex on a WBA resin, either the pH of the cyanide leach liquor should be adjusted or a resin with a pKa value greater than 10 must be used. A number of such resins has been synthesized (Green and Potgeiter, 1984) containing groups such as imidazoline or guanidine. However, these resins showed short life time due to chemical degradation.

It should be noted that conventional styrenic WBA resins have a certain strong base capacity and therefore they have a certain capacity for gold recovery with eventually different selectivity for gold than SBA resins. This high selectivity for gold can be enhanced by adjusting the ratio of strong to weak base functional groups of the resin. For example, a di- or a trivalent cyanide complex ($Zn(CN)_4^{2-}$ or $Cu(CN)_4^{3-}$) require two or three adjacent strong base functional groups to be able to be fixed by the resin while gold cyanide, $Au(CN)_2^-$ or $Au(CN)_4^-$ require only one. Therefore a resin having strong base sites situated at a long distance apart would fix preferentially monovalent cyanide complexes than polyvalent complexes. Such resins were used in the former Soviet Union countries (AM-2B) having 16-20% strong base groups (Lukey *et al*, 1998). Because of the presence of strong base groups, elution from these styrenic weak-base resins is not achieved with caustic solution.

The selectivity of anion exchange resins for metal cyanide complexes is very high to the extent that elution is very difficult. This high selectivity is believed to be due to the polarizability of the metal complex by interacting with fixed functional group of the resin.

Elution of strong base resins can be effected by treating the resin with zinc cyanide, thiocyanate or thiourea.
The basis for using zinc cyanide is that the resin has higher affinity for zinc cyanide than for gold cyanide:

$$2\ R\text{-}N^+R_3[Au(CN)_2]^- + [Zn(CN)_4]^{2-} \Leftrightarrow$$
$$(R\text{-}N^+R_3)_2[Zn(CN)_4]^{2-} + 2\ [Au(CN)_2]^-$$

After elution, the zinc cyanide is removed from the resin with dilute H_2SO_4:

$$(R\text{-}N^+R_3)_2[Zn(CN)_4]^{2-} + 2\ H_2SO_4 \longrightarrow (R\text{-}N^+R_3)_2SO_4 + 4\ HCN + ZnSO_4$$

This elution and regeneration scheme is based on the fact that $[Zn(CN)_4]^{2-}$ not only is strongly fixed on the SBA exchange resins but also it is the weakest of the weak-acid dissociables (WAD, page 155)) cyanide complexes and dissociates easily with H_2SO_4.

Thiocyanate has also higher affinity for the resin than gold cyanide:

$R\text{-}N^+R_3[Au(CN)_2]^- + SCN^- \leftrightarrows R\text{-}N^+R_3SCN^- + [Au(CN)_2]^-$

Thiocyanate is subsequently removed from the resin with 0.5-1 M ferric sulfate to form a cationic complex $Fe(SCN)_2^+$ not fixed by an anion exchanger. The SCN- are finally recovered with NaOH where iron precipitates out as iron hydroxide:

$Fe(SCN)_2^+ + 3\ OH^- \longrightarrow Fe(OH)_3 + 2\ SCN^-$

Thiourea elution is mostly used in the former Soviet Union RIP plants. It is based on the formation of a cationic complex of gold with thiourea which is released by the resin:

$R\text{-}N^+R_3[Au(CN)_2]^- + 2\ CS(NH_2)_2 + 2\ H_2SO_4 \leftrightarrows$

$R\text{-}N^+R_3HSO_4^- + 2\ HCN + [Au(CS(NH_2)_2)_2]^+HSO_4^-$

At the Golden Jubilee mine in S.Africa, the PLS contained high levels of organics (humic and fulvic acids) which fouled severely the C*. The plant switched in 1988 from CIP to RIP using a SBA resin. Elution initially was with thiourea but this became uneconomical due to solution fouling. They switched then to zinc cyanide elution which was satisfactory. The mine shut down in 1994.

Except these elution techniques which involve aqueous solutions, it has been found that the presence of organic solvents enhance the acid elution of gold cyanide from SBA resins, while aqueous acidic solutions did not elute gold cyanides (see also p.251). The solvent may enhance the elution of gold cyanides by either weakening the attraction of gold cyanide by the quaternary ammonium group of the resin, or by affecting the equilibrium with the result that the distribution of the complex between resin and solution changes.

Ores that contain carbonaceous or other materials that are "preg robbers", that is, as gold goes into solution, it is re-adsorbed by these materials thus depleting the solution from soluble gold, are called "preg-robbing" ores. In order to compete with the preg-robbing and to enhance leaching of gold, ion exchange resins can be used in the leaching solution (resin-in-leach, RIL) where they fix gold before it is absorbed by the preg-robbers (Schmitz *et al*, 2001; Goodall *et al*, 2005). RIL are in general more efficient than carbon-in-leach (CIL) for preg-robbing materials since resins compete more favorably preg-robbers like carbonaceous materials. The Avocet's Penjom mine in Malaysia, recovers gold from carbonaceous oxidized ore. The mine switched from CIL to RIL using Dowex® XZ 91419, whereby gold recovery increased from 60% to 85%. This installation operates since 1999. Elution is done with acidic thiourea solution.

Strong base resins have high selectivity for all cyanide complexes especially for high valence ones. Selective resins have been developed that are more selective for gold and silver cyanides than for cyanides of other metals like Cu, Ni, Zn, Co and Fe which are di- or polyvalent (Green et al, 1992). These resins are SBA resins having alkyl groups with more than three carbon atoms, n-butyl being preferred. Such a resin is the Dowex®-Minix® SBA resin developed jointly by The Dow Chemical and Mintek of South Africa. The gold and silver cyanide complexes as well as $Cu(CN)_2^-$ or $Ni(CN)_3^-$, being monovalent are preferred by these resins to the di- or polyvalent base metal cyanide complexes probably for steric reasons. Elution is carried out in two steps (van Deventer et al, 2000). In the first, an acid wash elutes the base metals. In the second, thiourea (1M) and H_2SO_4 (0.5M) at 60°C elutes gold. The resin is found under the commercial name of Dowex® XZ-91419.

A comparable situation represents the increased selectivity for the monovalent NO_3^- anions versus the divalent SO_4^{2-} with the nitrate-selective quaternary triethylamine anion exchangers. In fact, these resins have been used to recover gold as well (Schoeman *et al*, 2013, page 119).

Another selective resin developed is the AuRIX®100 (Cognis Corp, Tucson, AZ), having guanidine functional groups.

$$\underset{H_2N}{\overset{NH}{\underset{\|}{C}}}\diagdown NH_2 + H_2O \rightleftharpoons \underset{H_2N}{\overset{NH_2}{\underset{+}{C}}}\diagdown NH_2 + OH^-$$

Guanidine has a pKa about 12.5 and is therefore a moderately strong base. In fact in water exisis as guanidinium ion. The resin can therefore operate at pH<11 and fix gold from cyanide leach liquors. Elution of this resin is done with 0.5 M NaOH, 0.5 M sodium benzoate and 100 ppm NaCN at 55-60°C. The sodium benzoate improves the kinetics of elution but the total number of BV's needed for complete elution is the same as without sodium benzoate (Kotze *et al*, 2005).

As mentioned earlier, WBA resins containing an appropriate level of strong base groups have a certain selectivity for monovalent metal cyanides over polyvalent due to the fact that at the high pH of the PLS the weak base functionality is not ionized and therefore can not fix any anions while the strong base groups can. By adjusting the level of the strong base groups so that the distance between two adjacent strong base groups is far enough to favor monovalent over polyvalent anions.

One such WBA resin with mixed strong and weak base functional groups is used in the Muruntau mine in Uzbekistan. Elution is probably done in three steps (Marsden and House, 2006, p. 351)

a) acid to remove Zn, Ni b)NH4OH/NH4NO3 to remove Cu c) thiourea-EW for gold and silver

Gold recovery from cyanide leach PSL in copper containing ores presents difficulties to C* because copper cyanides are loaded on the carbon along with gold, thus decreasing the capacity for gold. At the pH and cyanide concentrations usually found in leaching, (pH>10 and excess of free CN^-) the prevailing species of copper cyanide are $Cu(I)(CN)_3^{2-}$ and $Cu(I)(CN)_4^{3-}$. If cyanide concentration and pH are low, as in heap leaching, then the monovalent species $Cu(I)(CN)_2^-$ prevails (van Deventer *et al*, 2000).

The selectivity of C* for copper cyanides is (van Deventer *et al*, 2014):

$$Cu(I)(CN)_4^{3-} < Cu(I)(CN)_3^{2-} < Cu(I)(CN)_2^-$$

In CIP or CIL operations therefore, the CN- concentration should be high in order to minimize the formation of $Cu(I)(CN)_2^-$.

Gold recovery from copper-gold PSL is possible with selective resins such as Dowex® XZ 91419, AuRIX®100 and other commercial resins like Purogold® S992 (van Deventer *et al*, 2014). Even with the above selective resins however, Cu is also removed along with gold with the same selectivity sequence as with C* indicated above. Split elution is possible, for example, with Purogold S992 separation of Cu from Au was done with two-steps elution, first with low NaOH concentration + NaCN at ambient temperature to elute Cu then with higher NaOH concentration + NaCN at 60°C to elute gold. As mentioned above, in those cases, elution with Dowex® XZ-91419 is carried out in two steps. In the first, an acid wash elutes the base metals. In the second, thiourea (1M) and H_2SO_4 (0.5M) at 60°C elutes gold.

The Gedabek mine in Azerbaijan is a copper-gold mine. Initially (2009) using heap leaching, where resins operate in RIS, since 2013 it uses agitation leaching where the resin operates in RIP installation. The resin used is Dowex® XZ91419. The RIS fixed bed was working at 6 BV/h, 4 parallel lines, 6.5 m3 resin in each (E&MJ News, 2010). Elution was done with 0.2 H2SO4 and 1M thiourea at 50°C.

Poisoning of the resins by cobalt cyanide is a problem encountered frequently. The responsible complex for this poisoning is the pentacyanide mono aquo species (Fleming and Hancock, 1979) which at pH<9 polymerizes inside the resin beads and then it can not be eluted out.

It should be pointed out here that since SBA resins are not selective for gold but can also fix all metal cyanide complexes, they can be used to remove metal cyanides from tailings, after the removal of gold by selective activated carbon or resins (Nesbitt and Petersen, 1994). An acrylic SBA resin Amberlite® IRA958 has given good metal cyanides removal in fluidized bed column. Elution was done with 2N NaCl regenerant which removed easily the divalent Ni(CN)42- and only partially the trivalent $Fe(CN)_6^{3-}$, $Cu(I)(CN)_4^{3-}$ and $Co(III)(CN)_6^{3-}$ complexes.

The equipment used in gold recovery includes fixed beds in cases of clear solutions: carbon-in-solution (CIS) or resin-in-solution (RIS) (for filtered or by heap leaching), fluidized beds for light pulps (heap leaching) or contactors treating pulp with up to 50% solids: carbon-in-pulp (CIP) or resin-in-pulp (RIP).

From cyanide leach solution, CIP is the prevalent technology in S.Africa while anion exchange resins in RIP systems are used in the former Soviet Union countries. Nevertheless, resins have also been evaluated by several companies and institutes in S.Africa and compared RIP to CIP technologies (Fleming and Cromberge, 1984).

Generally speaking, the differences between activated carbon, C^*, and resins can be summarized as follows:
- C^* is less expensive than resins
- resins have higher operating capacities than C^* (although this depends on the composition of the PLS).
- C^* has higher selectivity of gold cyanide than resins. However, resins are more versatile because they can be customized to improve selectivity and loading capacity
- Resins have higher resistance to fouling with organic matter (kerosene from the SX plant in uranium-and-gold mines, floatation chemicals, machine oils, humic and fulvic materials etc) and also to plugging the pores with fine particular calcine resulted from roasted pyrite, clay particles, hematite and shales.
- Resins extract more gold in RIL in preg-robbing ores.
- Simpler elution (<60°C and atmospheric pressure)

- Requirement of periodic thermal reactivation of C* to remove organics which results in high capital and operating costs and is responsible for up to 50% of the carbon losses.
- Slow rate of adsorption of $Au(CN)_2^{2-}$ on carbon (unaccessibility of micropores to $Au(CN)_2^{2-}$)
- Higher carbon losses due to abrasion, fracturing and burn-off in the kiln
- Resins can extract other metal cyanides, thus producing environmentally acceptable tailings and allowing the recycling excess cyanide.
- High concentrations of resins in adsorption container (20-30% by volume)

Non-cyanide leaching has been recently reviewed (Aylmore, 2016).
Gold recovery from thiosulfate leach liquors can be achieved with SBA exchange resins (O'Malley, 2002; Alfaro and Frenay, 2004):

$$3\ R\text{-}N^+R_3\ X^- + [Au(S_2O_3)]^{3-} \leftrightarrows (R\text{-}N^+R_3)_3[Au(S_2O_3)]^{3-} + X^-$$

Elution is effectuated with thiocyanate:

$$(R\text{-}N^+R_3)_3[Au(S_2O_3)]^{3-} + 3\ SCN^- \longrightarrow 3\ R\text{-}N^+R_3SCN^- + [Au(S_2O_3)]^{3-}$$

Copper is also loaded on the resin as $[Cu(S_2O_3)_3]^{5-}$ and is selectively eluted with thiosulfate prior to the elution of gold.

The resin after elution is regenerated not with an acid like in the $Zn(CN)_4^{2-}$, because thiocyanate decomposes to give elemental sulphur which poisons the resin, but with ferric sulfate which forms a cationic complex with thiocyanate, $Fe(SCN)^{2+}$ (Fleming, 1985).

From thiourea leach liquors, since the thiourea complex is a cation, cation exchange resins are employed (Nakahiro *et al*, 1992). The resin studied was Amberlite® 200, a macroreticular SAC resin made by Rohm and Haas Company (now The Dow Chemical Company). However these resins are not selective enough for gold and load other base cations present in the solution.

Recovery of gold from acidic chloride leach liquors has been achieved with the acrylic neutral adsorbent Amberlite ® XAD7 (of Rohm and Haas, now The Dow Chemical Company) (Harris and White, 2012). Elution was performed with diluted HCl solution. The eluate was then treated with $FeCl_2$ whereby gold precipitated as Au° and the produced $FeCl_3$ was recycled in the leaching step:

$$HAuCl_4 + FeCl_2 \longrightarrow Au° + FeCl_3 + HCl$$

Both strong and weak base resins can be used to fix gold from acidic chloride leach liquors. Stripping can be done with alkaline solutions like thiosulfate or cyanide solutions.
Gold from acidic chloride liquors has been achieved with the WBA resin Lewatit® MP-64 (from Lanxess) (Alguacil *et al*, 2005). Elution of gold was done with acidic thiourea solutions.

From acidic thiocyanate leach liquors, solvent extraction or ion exchange has been used (Wan and Levier, 2008). Purolite^{TM} 600, a gel type SBA resin was employed in batch experiments.

Cyanides removal from wastes

Cyanide is a strong ligand and forms easily complexes with most metals. Because of that, it is used widely in metal finishing operations and in metal leaching from ores, mainly in gold hydrometallurgy.
In the wastes of these industries cyanides can be found either as free cyanide, CN^-, or bound on metals forming anionic complexes. The metal cyanide complexes are classified in weak-acid dissociables (WAD) which are relatively unstable complexes with metals such as Cd, Zn, Ni and Cu (partially), and strong-acid dissociables (SAD) with metals such as Fe, Co(III), Au and Ag.
Cyanides recovery from the tailings of gold cyanide leaching plants can be achieved by direct acidification and volatilization of the formed HCN followed by scrubbing with alkali (AVR process, **A**cidify, **V**olatilize, **R**eneutralize).
Indirect cyanides recovery is possible using SBA exchange resins
(Fleming, 2003 and 2005).
SBA exchange resins have high affinity for metal cyanide complexes. The loading reaction can be presented for the case of $Zn(CN)_4^{2-}$ by:

$$(RN^+R_3)_2SO_4^{2-} + Zn(CN)_4^{2-} \rightarrow (RN^+R_3)_2 Zn(CN)_4^{2-} + SO_4^{2-}$$

Since the resin has little affinity for free CN^-, the CN^- should leak through. To prevent this, a salt of Zn^{2+} is be added to the tailings in a pre-complexation reactor to convert free CN^- to the complexed form. In this way, all cyanides from the tailings are fixed on the SBA resin.

Elution is done with H_2SO_4 which corresponds to the acidification of the AVR process:

$(RN^+R_3)_2 Zn(CN)_4^{2-} + 2 H_2SO_4 \rightarrow (RN^+R_3)_2SO_4^{2-} + ZnSO_4 + 4 HCN$

The separation of the HCN from the eluate is done by the AVR process and the mother liquor containing $ZnSO_4$ is recycled to the pre-complexation reactor.

With gold ores containing high levels of copper, processes have been developed to recover cyanides (Augment process). The resin contains intentionally CuCN precipitate inside the resin beads which acts as an absorber for free and complexed cyanides (Fleming, 2005; Marsden and House, 2006; Dai et al, 2012). Again a pre-complexation with copper can be done here to convert free CN^- to the complexed form:

$(RN^+R_3)_2SO_4^{2-} \cdot CuCN + Cu(CN)_3^{2-} \rightarrow 2 (RN^+R_3)Cu(CN)_2^- + SO_4^{2-}$

Elution is performed with a concentrated copper cyanide solution with a molar ratio of cyanide to copper of 4:1 to 5:1. This removes about 50% of the copper and 25-30% of the cyanide. The elution reaction is represented by:

$2 (RN^+R_3)Cu(CN)_2^- + Cu(CN)_3^{2-} + 2 CN^- \leftrightarrows$
$(RN^+R_3)_2Cu(CN)_3^{2-} + 2 Cu(CN)_3^{2-}$

In the regeneration step, the acid intentionally precipitates CuCN in the pores of the resin:

$(RN^+R_3)_2Cu(CN)_3^{2-} + H_2SO_4 \rightarrow (RN^+R_3)_2SO_4^{2-} \cdot CuCN + 2HCN$

which is then ready for loading.

Cyanides can be removed from metal finishing wastes with ion exchange using the system SAC → WBA → SBA (see page 117).
The SAC resin operates in the H^+ form and removes all free cations including those forming WAD cyanide complexes, Cd, Zn, Ni and Cu (partially). The WBA resin operates in the protonated form, for example in the Cl^- form, and fixes the SAD metal cyanide complexes such as $Fe(CN)_6^{4-}$ and $Fe(CN)_6^{3-}$. The SBA resin operates in the OH^- form and removes HCN which is a weak acid and can not be removed by the WBA resin. Gel or MR type SAC resins can be used. Operating capacities obtained vary at 0.8-1.0 eq/L_R using 6-10% HCl regenerant at a level of 100-120 g/L_R or 10% H_2SO_4 at a level of 150 g/L_R. The WBA resin is converted to the Cl^- form using 3 BV of a 3% HCl solution. The SBA resin can be gel or MR type. Type 2 gel SBA resins are frequently employed. Regeneration is achieved with 4% NaOH at 100 g/L_R. Operating capacities are ln the range of 0.3-0.4 eq/L_R.

It is interesting to note that the chemistry involved in copper cyanide removal with IER is known since more than 50 years (Weiner, 1963): with a system containing SAC(H^+) → WBA → SBA(OH^-) resins, potassium copper cyanide is first dissociated partially on the SAC (H^+) resin where copper along with K^+ are fixed and HCN is formed. Another part of the complex, converted to the acid form, is fixed on the WBA resin while a third part precipitates in the SAC resin as cuprous cyanide, CuCN. The HCN is fixed on the SBA(OH⁻) resin. Regeneration of the anion exchange resins is done with NaOH starting from the SBA resin where NaCN is formed. The regenerant goes through the WBA resin to the SAC resin where the NaCN of the spent regenerant dissolves the CuCN precipitate.

In another process (Goldblatt, 1959) two SBA resin columns were employed in series, the first of which was in the SO_4^{2-} form while the second was also in the SO_4^{2-} form but in addition, "conditioned" to incorporate CuCN precipitate in the resin. Copper cyanide complexes were fixed on the first column while free cyanides on the second column.

Platinum Group Metals

From the various hydrometallurgical flow sheets, PGM can be recovered with ion exchange either from acidic leachates where they are found as chlocomplexes, or from cyanide leachates (originating from gold extraction plants).

PGM recovery from acid leachates with ion exchange can be done with:
* chelating resins or MRT resins
* anion exchange resins, either strong or weak base
* reducing resins having boron hydride functional groups

Of these, chelating resins, such as thiourea resins, fix well PGM but eventually are difficult to elute and burning them to recover the precious metals may increase the cost of the operation. Otherwise, elution is possible with acidic thiourea (Schmuckler,1969). Reducing resins containing boron hydride functionality were produced by Rohm and Haas Company under the trade name of Amborane® but today are no longer commercially available.

If gold is present, the first step is the separation of gold from the PGM. This can be achieved with an acrylic type polymeric adsorbent such as Amberlite® XAD7 by Rohm and Haas Company (today The Dow Chemical Company) (Edwards *et al*, 1976). The separation is based on the interaction of the metal chlorocomplexes with compounds bearing carbon-oxygen bond (like ethers or dibutylcarbitol). The distribution coefficient between the aqueous medium and the organic phase depends on the ratio of the charge to the size of the anion. Thus, Au(III) complex $[AuCl_4]^-$ should be extracted more easily than the PGM chlorocomplexes. Hydrochloric acid concentrations found to favor the capacity of the adsorbent for gold was between 2N and 6N. Elution was done with mixed acetone and 4 M HCl solution. Gold is recovered from the spent eluent by precipitation with a reducing agent (SO_2 or formic acid).

The key parameters for efficient leaching of PGM are the Cl^- concentration and the redox potential (Rumpold and Antrekowitsch, 2012; Lillkung et al, 2013). Each platinum group metal has its own redox potential and in fact, this can be used to separate the PGM from each other. Leaching is done with hydrochloric acid and an oxidant such as chlorine or hydrogen peroxide.

A flow sheet developed for the production of PGM from low-sulfides platinum ores (Tatarnikov *et al*, 2004) consists of an autoclave oxidative leaching to dissolve non-ferrous metals, followed by roasting and hydrochlorination to put the PGM in solution as chloride complexes. The PGM were recovered with a

WBA resin containing primary, secondary and tertiary amine functional groups. The redox potential was <800 mV to avoid resin oxidation. Flow rate was 2 BV/h. The treated solution was used to prepare the hydrochlorination solution so that any residual PGM were recycled. The loaded resins were not eluted but they were burnt to recover the precious metals.

In another work (Green *et al*, 2004) PGM were recovered from HCl/Cl$_2$ leachates with a poly-4-vinylpyridine resin (Reillex® HP425) which was subsequently eluted first with water to desorb any base metals (by converting the anionic complexes to metal cations) followed by acidic thiourea elution to elute the PGM.

From cyanide leach solutions containing gold, PGM and base metals, the recovery of gold was achieved using the Dowex®-Minix® resin bearing tributyl ammonium functional groups and which removed selectively gold, while the nitrate-selective resin Amberlite® PWA5 bearing triethyl ammonium functional groups removed Au(CN)$_2^-$, Pd(CN)$_4^{2-}$ and Pt(CN)$_4^{2-}$ as well as Zn(CN)$_4^{2-}$ and Ni(CN)$_4^{2-}$ were removed in presence of base metals Cu(CN)$_4^{3-}$, Cu(CN)$_3^{2-}$ and Fe(CN)$_6^{4-}$ (Schoeman *et al*, 2013).

In some cases where platinum is used as catalyst in alkaline media, by corrosion, Pt dissolves and forms [Pt(OH)$_6$]$^{2-}$. In this case, an anion exchange resin can be tried to recover it.

Base metals

RIP is a potentially interesting technology for the recovery of base metals from dilute solutions and pulps that are difficult to settle or filtered and from various solid residues.
Nickel and cobalt are extracted from lateritic ores by pressure acid leaching (PAL) process. After dissolving nickel and cobalt with acid at high temperature and pressure, the metal-containing liquid is separated from the solids by filtration or by settling in a series of thickeners. Nickel and cobalt are subsequently recovered from the clear leach solution by techniques such as precipitation, SX and electrowinning. A simplified flow sheet is shown below:

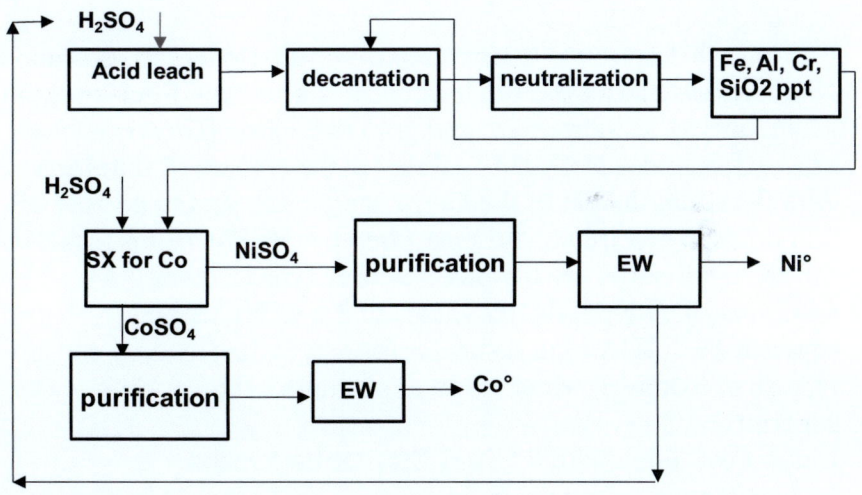

Ion exchange can be used as follows:
a) recover Ni and Co from counter-curent decantation (CCD) or thickener underflow with resin-in-pulp (RIP) technology.

b) recover Ni and Co directly from leached pulp without solids-liquid separation, using RIP technology
c) purify cobalt electrolyte

These possibilities are discussed below:
a) Nickel and cobalt recovery from the tailings pulp is achieved with a IDA type resin (Zainol, 2005). The feed pulp is first neutralized to a pH of 4.5 in order precipitate iron, previously oxidized to Fe(III), chromium and aluminum and facilitate Ni and Co recovery. The loaded resin with Ni and Co is eluted with 1M H_2SO_4 solution. Any Fe^{3+}, Cr^{3+} and Al^{3+} loaded on the resin need higher H_2SO_4 concentrations to be eluted because they are stronger held by the resin. With this selective elution, Ni and Co are separated from the impurities.

b) A possible option is to use a resin-in-pulp system to recover Ni and Co directly from the leach solution without thickening, in a similar way as in uranium and gold extraction (Duyvesteyn *et al*, 2002; Mendes *et al*, 2005). Prior to the contact of the slurry with the resin, the pH of the slurry can be adjusted to a value of 3 in order to precipitate out metal impurities. The resins tested for this application are bis-picolylamine type (Dowex® M-4195) and IDA type. At pH values of 3.5 or higher and in absence of Fe^{3+}, IDA type resins performed better (loading capacity, rate of loading), while at lower pH values the BPA resin was superior.
Elution was done with 0.5 N H_2SO_4 for both resins.
In a analogous approach, resin-in-leach (RIL) technology was suggested to recover nickel and cobalt (Mendes, 2009). The metal values are leached with acids and simultaneously ion exchange resin is added to the pulp. The metals are loaded on the resin as they are solubilized. The pH of the pulp is adjusted to 1-

3. Trivalent iron and copper are simultaneously eliminated from the pulp by adding magnesium, aluminum and iron. BPA type resins (Dowex® M-4195) were found more suitable. Step-wise elution was performed: first, iron was eluted with acidified (pH=2) water. Then copper is eluted with $(NH_4)_2SO_4/NH_4OH$. Finally, Ni and Co are eluted with acid such as H_2SO_4. Nickel and cobalt can then be recovered from the spent eluent by conventional methods such as precipitation, SX or membranes.

c) Cobalt advance electrolyte was purified by removing copper and zinc (Strong and Henry, 1976). Copper was removed using an IDA type resin (Lewatit® TP207) while zinc was removed using a solvent impregnated resin (SIR) impregnated with D2EHPA specially made by Bayer (today Lanxess). This product is available today by Lanxess under the name Lewatit® VP OC 1026.

The selectivity of IDA resins is much higher for Cu^{2+} compared to Co^{2+} and in fact, copper can be removed with the resin in the H^+ form. Two columns in series were used with a copper concentration at the exit of the polishing column of less than 0.1 mg/L with a feed concentration of 40 g/L Co^{2+} and 11.7 mg/L of Cu^{2+}. Specific flow rate was 5-20 BV/h. elution was done with 10% w/w H_2SO_4 at 2 BV/h. The elution curves showed that the bulk of the cobalt was eluted before copper and this allows to separate to a certain extent the two metals.

Zinc was removed with the SIR from the cobalt electrolyte after the removal of copper. pH was adjusted to 5 and specific flow rate was 10 BV/h. Zinc concentration at the exit of the polishing column was less than 0.1 mg/L after 50 hours cycle, with a feed containing 43 mg/L Zn. Elution was done with 4 BV of 5% w/w H_2SO_4 at 2 BV/h. Both, Co and Zn were eluted together and no separation was possible.

Copper and zinc removal from cobalt electrolyte has also been investigated using an AMP type resin in the Na form, either to remove both metals with the IER or use SX to remove Zn followed by AMP resin to remove copper (Kotze, 2012; Taut *et al*, 2013).

The removal of low levels of nickel from cobalt sulphate electrolyte using ion exchange was achieved with BPA resins (Dowex® M-4195). This resin fixes both, cobalt and nickel but it has a higher selectivity for nickel over cobalt. For example, with a feed solution containing 60 g/L cobalt, 300 mg/L nickel, 50 g/L sodium sulphate7 g/L magnesium and 400 mg/L calcium (as sulphate) at 5 BV/h and 60°C the resin loadings were close to 23 g/L_R cobalt and 5.9 g/L_R nickel (Bailey et al, 2005). Elution was performed using split elution technique. The majority of cobalt was eluted with 3-4 BV of 25-30 g/L H_2SO_4 with little nickel eluted under these conditions. Nickel was eluted with 2-3 BV of 100-150 g/L H_2SO_4. These elution conditions reflect in fact the selectivity difference of the resin for nickel and cobalt.

Manganese is found frequently with other transition metals as MnO_2. Many of the processes to treat manganese minerals utilize acidic concentrated chloride solutions. Typical metal concentrations in such solutions are Mn^{2+} 5 g/L, Fe^{3+} 5 g/L, Cu^{2+}, Ni^{2+}, Co^{2+} and Pb^{2+} 50 mg/L each. In order to separate the metal impurities from manganese, ion exchange has been evaluated. It was found (Diniz et al, 2002) that BPA resins such as Dowex® M-4195 presented a potential interest because it showed high affinity at high Cl^- concentrations (3.6 M) for all metal impurities while it showed low affinity for Mn^{2+}.

Cobalt was separated and recovered from copper recycling leach liquors using the BPA resin Dowex® XFS-4195 (Jeffers, 1985), today named Dowex® M-4195. The feed solution contained 30 ppm Co, 30 ppm Ni, 60 ppm Cu, 2 g/L Fe, 200 ppm Zn and 4.5 g/L Al and had a pH of 3. XFS-4195 removed Co, Ni, Cu and Zn and much less Fe. By eluting with 5 BV of 50 g/L H_2SO_4, all metals were eluted except Cu. Copper was eluted with 3 BV of 2N NH_4OH. The metal impurities from the acidic eluate were removed using SX techniques: D2EHPA removed Zn, Al and most of Fe at a pH of 2.5, then Cyanex 272 removed the rest of Fe. Finally, the remaining Co and Ni were separated with Cyanex 272 at a pH of 5.

Zinc recovery from various solid effluent streams has been evaluated using RIP technology (Kotze, 2012). An IDA type resin was employed due to its high capacity and selectivity for Zn^{2+} and easy elution with H_2SO_4.
The purification of a leach solution containing 32 g/L Cu and 73 g/L Zn originating from a copper smelter flue gas leaching, was achieved with a HPPA type resin (Dowex® XFS-43084) (Jha et al, 2001). After loading the resin, elution with 1.5 M H_2SO_4 produced a copper rich solution containing 18-20 g/L copper and 3-4 g/L zinc while the raffinate stream contained 62-64 g/L zinc and 6-7 g/L copper.

In copper recovery, an attractive technology is SX-EW. If IX is envisaged, there is a need for the reduction of the cost in the post leach solid washing and solid-liquid separation equipment. The use of continuous ion exchange (CIX) in fluidized beds technology, as in uranium recovery, is therefore an attractive alternative especially with low grade oxide sources (Naden and

Willey, 1974). The use of chelating resins of the IDA type has been evaluated in CIX systems. The low selectivity however of these resins for Cu^{2+} over Fe^{3+} leads to the use of SX to extract Cu^{2+} from the H_2SO_4 eluates. Reduction of Fe^{3+} to Fe^{2+} with SO_2 is an option to improve this CIX-SX process.

The use of RIP technology has been evaluated for the recovery of copper in CCD underflows (Greager *et al*, 2001). Selective chelating resins were evaluated. The selectivity sequence was $Fe^{3+}>Cu^{2+}>Fe^{2+}>Ca^{2+}>Mg^{2+}>Na^+$. Consequently, before recovering copper, Fe^{3+} was removed by hydrolytic precipitation as iron hydroxide. The effect of Cu^{2+} concentration in the feed (0.5 to 4 g/L) and the effect of pulp pH (2.5 or 3.0) were studied. Pilot plant studies were conducted in order to dimension the plant. Four stages RIP plant were found adequate to bring Cu^{2+} concentration from 0.5-4 g/L down to <0.05 g/L. Elution was done with H_2SO_4.

Recovery of copper and cobalt from a tailings dump using RIP technology and an IDA type resin was described (Yahorava and Kotze, 2011). The two metals, Cu and Co, were recovered simultaneously with one RIP unit (rather than two units, one for Cu and one for Co due to the pH difference in the absorption by the resin). Although Cu started precipitating at the pH of 4.5 that cobalt is fixed on the IDA resin, it was possible to recover both metals.

An ISEP[TM] system has been studied for the recovery of copper in presence of iron (Rossiter and Carey, 1998). A special chelating resin from Dow Chemical was employed showing high affinity for Cu^{2+} over Fe^{+3} (Dowex XFS-43084, a HPPA resin, see p. 45-46). The characteristic of this resin in comparison with the

BPA analogue Dowex® M 4195 is that it has a higher selectivity for Cu over iron and also that it can be eluted with lower concentration of H2SO4.

In copper electrorefining, impure copper is the anode of an electrolytic cell, pure metal the cathode, with the electrolyte solution that consists of $CuSO_4$ and H_2SO_4. As electric current passes through, copper dissolves from the anode and passes to the electrolyte solution while copper is deposited on the cathode. The electrolyte, which contains H_2SO_4 at 50 g/L or more, is recovered and reused as much as possible. However, the eventual impurities contaminate the electrolyte and deteriorate the quality of the refining. Two of the contaminants are antimony and bismuth. In order to remove Sb and Bi at this H_2SO_4 concentration with chelating resins seems difficult. Removal of antimony and bismuth from copper electrolyte however has been reported (Tadao and Yoshiaki, 1985) using a phenolic resin bearing methylphosphonic groups, Unicellex™ 3300 of Unitika Ltd. A styrene-DVB type resin could also be used, of the AMP type. Loading specific flow rate was 1-5 BV/h. Regeneration of the resins was done with concentrated HCl. This patent was subsequently improved by reducing the Fe^{3+} to Fe^{2+} in order to avoid poisoning of the resins by Fe^{3+} (Dreisinger and Leong, 1994). In a more recent patent (Riveros, 2013), because the resin was fouled by accumulation of Sb(V) while the Sb(III) elution was satisfactory, a regeneration with acidic thiourea was proposed.

Group IV: zirconium, hafnium. Group V: vanadium, niobium, tantalum. Group VI: molybdenum, tungsten. Group VII: rhenium,

Zirconium and Hafnium form compounds usually in the +4 state. They have similar properties and in general some contamination of Zr by Hf can be tolerated except for nuclear applications. Separation of Zr and Hf can be done with SX or fractional crystallization.
Early investigators reported IX processes to separate Zr and Hf (Gupta and Mukherjee, 1990). From oxychloride, nitrate and sulfate solutions, on a SAC resin. Selective elution to separate Zr from Hf.
From fluoride solutions, Zr and Hf form anionic complexes ZrF_6^{2-} and HfF_6^{2-}, and are fixed on SBA resins. Selective elution with HCl+HF

Tantalum and niobium are transition metals of the V group. Ta has the electronic structure $[Xe]4f^{14}5d^36s^2$ and is found in the +4 and +5 oxidation states, the most stable being the +5, Ta_2O_5. Niobium has the electronic structure $[Kr]4d^45s^1$ and has similar properties (Nb_2O_5). Ta_2O_5 dissolves in basic hydroxide solutions to form tantalates TaO_4^{3-}.
Tantalum ores often contain niobium. Extraction starts with a leaching with hydrofluoric acid and sulfuric or hydrochloric acid:

$$Ta_2O_5 + 14\ HF \longrightarrow 2\ H_2[TaF_7] + 5\ H_2O$$
$$Nb_2O_5 + 10\ HF \longrightarrow 2\ H_2[NbOF_5] + 3\ H_2O$$

In Cl⁻ and F⁻ media Ta and Nb form fluoride complexes TaF_7^{2-} and $NbOF_5^{2-}$ and they can be fixed on a SBA resin like Amberlite™ IRA400 of Rohm and Haas Company (Gupta and Mukherjee, 1990). From there they can be eluted selectively, first with F⁻ and Cl⁻ solutions to remove impurities, then with F⁻, NH_4^+ and Cl⁻ solutions at pH=1 to elute Nb and then by F⁻, NH_4^+ and Cl⁻ solutions at pH=5.5 to elute Ta.

Molybdenum is a transition element of the VI group with an electronic structure $[Kr]4d^55s^1$. The most stable oxidation states of molybdenum are the +4 (MoS_2) and +6 (MoO_3). MoO_3 is soluble in alkaline water forming molybdates MoO_4^{2-}. Molybdenum is extracted mainly from sulfide ore, molybdenite (MoS_2). Low-grade Mo sources, where ion exchange technology may be envisaged, include copper ores containing rhenium and molybdenum and uranium ores.
In one case, low grade MoS_2 sources were leached with sodium hypochlorite, NaClO (Cox and Schellinger, 1958). The leach liquor containing 3% NaClO were treated with SBA resins (Amberlite™ IRA400) in the Cl⁻ form:

$$2 \text{ R-Cl} + Na_2MoO_4 \leftrightarrows R_2MoO_4 + 2 \text{ NaCl}$$

Elution was done with 8% NaOH solution:

$$R_2MoO_4 + 2 \text{ NaOH} \leftrightarrows 2 \text{ R-OH} + Na_2MoO_4$$

The resin was converted back to the Cl form with HCl.
Today, the hypochlorite leaching is no longer used due to corrosive effects.

In acid leaching of uranium ores containing molybdenum it forms sulfate complexes, $[MoO_2(SO_4)_n]^{n-2n}$, in a similar manner as uranium, and it is fixed on anion exchange resins, even stronger than uranium thus fouling the resin. From SBA resins, Mo can be eluted with 2-10% NaOH or with 20-50% H_2SO_4. From alkaline leach of uranium ores, molybdenum is fixed on the SBA resin but is displaced by the tri-carbonate complex of uranium.

Rhenium is found in molybdenite (MoS_2), itself being a by-product of porphyry copper ore deposits. After flotation, the molybdenite concentrate is roasted to convert MoS_2 to MoO_3 and SO_2. During this roasting, Re is oxidized to the volatile Re_2O_7
which is recovered with scrubbing with water where it forms perrhenic acid, $HReO_4$. The flue gas effluent contains about 0.3-0.7 g/L Re, 3-6 g/L Mo, 1-2 g/L Fe, Zn and Cu and 50-120 g/L H_2SO_4. The other possibility to extract Re from molybdenite is by leaching with HNO_3. The leach liquors contain 10-40 mg/L Re, 8-15 g/L Mo, 2-5 g/L Fe, Zn and Cu, 100-150 g/L H_2SO_4 and 10-30 g/L HNO_3. Rhenium can be fixed on anion exchange resins along with molybdenum. WBA exchangers are better for Re removal because it is difficult to be eluted from SBA resins while from WBA resins it can easily be eluted with ammonia to be recovered by evaporation as ammonium perrhenate, NH_4ReO_4.

It is possible to recover rhenium from acidic solutions with a WBA resin, however, conventional WBA resins are not selective for Re and fix MoO_4^{2-} at the same time. In order to increase the selectivity of the WBA resins for Re over Mo, resins with higher degree of crosslinking were made, with the idea that high crosslinking density would slow down the diffusion of the bulkier molybdate polyanions compared to the perrhenate anions (Gedgagov and Nekhoroshev, 2011). The Diphonix® type resin Purolite® S957 (see p. 39) was found to be selective for Mo over Re (Mikhaylenko, 2011).

Purolite International Ltd has developed WBA resins specially made for rhenium processing (Mikhaylenko, 2011). Purolite® A170, a macroporous styrenic weak base anion exchange resin is recommended for high acidity solutions and Purolite® A172, a gel-type styrenic weak base anion exchange resin, recommended for high selectivity for Re over Mo.

Evaluation of Purolite® A170 and Purolite® A172 resins was performed in flue gas effluents as well as in leach liquors of molybdenite (Blokhin *et al*, 2011; Joo *et al,* 2012). It was found that Purolite® A172 had a better selectivity for Re over Mo while Purolite® A170 loaded both, Re and Mo. Elution with 6M NH_4OH was efficient for Purolite® A170 but for Purolite® A172 it was efficient only if the resin was loaded with high levels of Re. Treating flue gas effluents, Purolite® A172 showed a better loading capacity for Re than Purolite® A170 and because of the high Re loadings, elution with 6 M NH_4OH was feasible. In addition, a selective elution of Mo with saline solutions allowed the recovery of purer Re. Evaluation of the two Purolite® resins for molybdenite concentrates from HNO_3 leaching showed that the presence of NO_3^- decreased the capacity for Re. The low Re loadings on Purolite® A172 did not allow elution with ammonia and consequently only Purolite® A170 was found suitable.
Rhenium can also be found in small quantities in uranium ores from which, due to the high demand, can be recovered from

both, acid and alkaline leach liquors (Troshkina, 2011). Re is fixed, along with uranium, on SBA resins. Elution of Re from SBA resins is difficult. Re is eluted with selective eluants: 80-90 g/L NO_3^- plus 4-4.5% HNO_3 from acid leach processes. Re is subsequently recovered from the spent eluent with SX from which Re is back exctracted with ammonia to recover NH_4ReO_4. From alkaline leach processes, Re is eluted with 1-2 mol/L NH_4SCN plus 1% NH_4OH from which it is recovered by evaporation and crystallization.

The above mentioned Purolite® A170 and Purolite® A172 resins have been evaluated in recovering Re from copper smelters and from uranium deposits in Kazakhstan (Abisheva and Zagorodnyaya, 2011). Purolite® A170 fixed well Re from a sulphuric acid solution from a copper smelter containing 60 g/L H_2SO_4 and 13 mg/L Re. After 3000 BV of solution treated, the resin had shown a capacity of 27 g Re/L_R to 95% leakage. Elution was done with 3 M NH_4OH solution and more than 96% of the fixed Re was eluted with 4 BV of eluent. The Re concentration in the 4 BV of eluent was found to be 5.4 g/L, which allows the recovery of ammonium perrhenate by evaporation.

The resins AN-21 (a WBA resin), Purolite® A170 and Purolite® A172 were compared with a H_2SO_4 underground leaching of uranium deposits liquor. The feed acid leach pregnant solution contained 60 mg/L U and 0.48 mg/L Re. Two of the resins, AN-21 and Purolite® A170 showed similar loading figures for U and Re at the exhaustion point for Re: 26.7 g U/L_R and 1.67 g Re/L_R for AN-21 and 27.5 g U/L_R and 1.77 g Re/L_R for Purolite® A170. Purolite® A172 on the other hand showed a significantly higher Re capacity and lower U capacity: 8.73 g U/L_R and 6.74 g Re/L_R. Elution of AN-21 and Purolite® A170 were done first with 3 M NH_4OH at 50°C to elute Re followed by H_2SO_4 / NH_4NO_3 elution to recover uranium. On the other hand, it was not possible to elute Re with ammonia solution from Purolite® A172. Another approached studied was to place Purolite® A170 after the uranium resin, to treat the uranium barren solution.

In order to increase the Re concentration, the spent eluent was neutralized with H_2SO_4 and loaded again on the Purolite® A170 after which an ammonia elution gave an average Re concentration in the spent eluent of 4 g/L which allows to recover ammonium perrhenate by evaporation.

Vanadium forms oxoanions usually in the +5 state, the simplest of which is the orthovanadate VO_4^{3-}. There are many examples of vanadate ions one of which is metavanadate, $[VO_3]_n^{n-}$. In the V^{5+} state, it can be found also in the solution as cation (VO_2^+). In the +4 state vanadium is found as cation VO^{2+}.

Depending on the amount of oxidant used, vanadium can be present as quadravalent, VO^{2+} or pentavalent, VO_3^- or VO_4^{3-}. In the +5 state therefore, vanadium is fixed on the SBA resin along with uranium. From ores that contain both uranium and vanadium, in order not to contaminate uranium with vanadium, the redox potential can be adjusted so that vanadium is found in the +4 state as cation.

Vanadium can be recovered from acidic leach liquors using strong base anion exchangers (Zipperian and Raghavan, 1985). In alkaline leach of uranium from carnotite ores, VO_3^- and VO_4^{3-} can be fixed on the SBA resin. The operating capacity of the resin depends on the total carbonate and bicarbonate concentrations. It has been found that vanadates are fixed significantly better from a HCO_3^- solution than from a CO_3^{2-} solution probably due to the higher resin selectivity for CO_3^{2-} ions (divalent) over HCO_3^- ions (monovalent). Therefore, if both uranium and vanadates are found in the pregnant solution, the absorption of vanadates is minimized if the pH of the solution is higher than 10. Under these conditions and provided that the cycle goes to the saturation of the resin, the vanadium fixed can be less than 1% of the uranium fixed on the resin.

Molybdenum and vanadium can be recovered from spent catalysts by alkaline leaching (Ackermann et al, 1993). After adjusting the pH to neutral in order to remove by precipitation alumina, phosphorous and arsenic, the pH is brought to about 4-5 and the solution is passed through a WBA resin in SO_4^{2-} form where both molybdates and vanadates are fixed. Elution is done first with H_2SO_4 acid containing a reducing agent (SO_2). In this step vanadium is eluted by conversion of VO_4^{3-} (V^{5+}) to VO^{2+} (V^{4+}) which as cation is not retained by the anion exchange resin. In a second step, molybdenum is eluted with ammonia.

Vanadium can be recovered from ammonium metavanadate solutions with SBA resins (Naumann, 1968). The metavanadate solutions were treated by a SBA resin (like Amberlite^{TM} IRA402 of Rohm and Haas Company) from which was eluted with NH_4Cl.

The Bayer liquor in the alumina production usually contains V_2O_5 present in bauxite. There are three ways to extract vanadium from the Bayer liquor: crysrallization, SX and IX. One IX technique (Zhao et al, 2010) consists in precipitating vanadium from the Bayer liquor, leaching with $NaHCO_3$, removing V from the leach liquor with a SBA exchange resin, eluting with 3 M NaOH and recover vanadium from the eluate by precipitation.

Molybdenum and vanadium can be recovered from sulfate acidic solutions of blue sludge (Guo and Shen, 2015). Molybdenum ($Mo_7O_{21}(OH)_3^{3-}$) and vanadium (VO_2^+) were fixed on a IDA type chelating resin (Amberlite™ IRC748) at a pH of 2 from which the two metals were separated by selective elution.

Tungsten is a transition element of the VI group, as molybdenum with which has similar properties. It can be recovered from ores either by alkaline leaching or by acid leaching. Tungsten at neutral pH comes either as orthotungstate, WO_4^{2-} or as paratungstate, $HW_6O_{21}^{5-}$. At acidic pH it comes as metatungstate, $H_2W_{12}O_{40}$. Strong base anion resins have been evaluated for the recovery of tungsten from dilute (<2g WO_3/L) solutions (Martins et al, 1984). As it can be easily understood, the resin capacity at low pH was found to be much higher than at higher pH, since the equivalent weight of W at low pH is higher. Regeneration was done with 2N NaCl solution.

Rare earths

Rare earth elements (REE) include scandium (Sc), yttrium (Y) and the 15 lanthanides series elements from lanthanum (La) to lutetium (Lu). From La to samarium, Sm, are called light REE, the rest heavy REE. More than 95% of the rare earth oxides are found in the minerals bastansite, monazite and xenotine. REE are also found in apatites and spent uranium solutions.

Scandium is found in uranium tailings, in waste tungsten sludges, in red mud in the alumina production or in laterite ores. The most common oxidation state is 3+.

Sc^{3+} can in principle be recovered from leach liquors using a SAC resin in the H form which however fixes all metal or other cation impurities. Subsequent Sc purification can be achieved by SX or selective elutions.

From tungsten sludges, one way to recover Sc^{3+} is with chelating IDA-type resins in the H form (Rourke et al, 1989). The sludges are leached with acid containing a reducing agent in order to reduce manganese (IV) and iron (III) to Mn(II) and Fe(II). The solution is then passed through the IDA type resin. Elution is first done with dilute HCl to remove easily desorbed metals and then with a chelating agent (diglycolic acid, HOOC-CH_2-O-CH_2-COOH) which elutes scandium. Scandium is finally recovered by precipitation.

After ore grinding, magnetic separation and other upgrading operations, the ores are leached with H_2SO_4 or HCl after which the lanthanides are separated from the solids by filtration and recovered as chlorides or hydroxides concentrates. These are further processed by calcining and acid leaching to produce rare earth solutions which are then treated with SX or ion exchange for separation and recovery of the individual REE.

The separation of rare earths with ion exchange is achieved by ion exchange chromatography. All lanthanides have the 3+ oxidation state. Therefore a sulfonic strong acid cation exchanger can be used to remove them from the solution (Kolodynska and Hubicki, 2012). The affinity of the resin however for the REE is about the same and consequently, separation of the REE is achieved with selective elution using complexing agents that form complexes with the REE. Such complexing agents include aminopolycarboxylic acids like EDTA and nitrilotriacetic acid (NTA), carboxylic acids, hydroxylic acids like citric and lactic acids and aminophosphonic acids. The SAC resin has higher affinity for the REE that forms the weakest complex. The separation efficiency therefore depends on the stability constant of the formed complexes. For example, with EDTA, the stability complex increases from La to Lu. Elution is rated in the order of decreasing atomic number.

Due to the low solubility of EDTA in acidic solutions, elution with EDTA with the resin residual capacity in the H^+ form does not take place efficiently. Therefore, the SAC resin is used not in the H^+ form but in some other form to load the REE. Cu^{2+} form resin was found to be suitable because the stability constant of the Cu-EDTA complex is higher than that of REE-EDTA. So, to separate REE, the SAC resin is first converted to the Cu^{2+} form by passing a $CuSO_4$ solution through. After that, the REE solution passes through and the REE are exchanged with the Cu^{2+} ions and are fixed on the resin. To separate the REE, an eluent such as $(NH_4^+)_4EDTA$ is passed through the resin. The REE have high affinity for EDTA and are displaced from the resin, leaving the NH_4^+ on the exchange sites. Each lanthanide has different affinity for EDTA so that each lanthanide is displaced and recovered separately. It is noted however that only one eluent cannot separate all REE from each other. Several complexing eluents should be used.

An alternative way is to have two SAC columns in series, the first to fix the REE and the second in the Cu^{2+} form (Powell *et al*, 1957). After fixing the REE in the first column in the NH_4^+ form in a conventional way, the EDTA eluent passes through and elutes the REE by forming the REE-EDTA complex. When the REE-EDTA complex enters the second column in the Cu^{2+} form, the Cu-EDTA complex is formed because it has a stability constant significantly higher than the REE-EDTA complexes and te REE displaces Cu^{2+} from the resin. In this way, by reabsorbing the REE in the second column, a better separation of the rare earths is achieved since the re-absorption of REE there is some separation due to the different selectivity of the REE-EDTA complexes for the resin. In the previous case described above, only one column is used with the resin in the Cu^{2+} form where the REE is re-absorbed at the lower part of the column.

Separation of thorium, Th, and uranium, U, is achieved with anion exchange resins based on the formation of anionic complexes of these two elements in acidic media while the lanthanides form less stable anionic complexes or form cationic complexes. Thus, Th forms anionic complexes in HNO_3 and U in HCl and are separated from the REE on an anion exchanger.

Separation of anionic complexes of rare earths using gel-type strong base anion exchangers is possible. REE show little tendency to form anionic complexes however in mixed water-organic solvents systems the formation stability of the complexes increases. Such solvents include lower alcohols, acetone, dioxane and acrtic acid. It was possible for example to use SBA exchange resins for separation of REE-HNO3-methanol systems with aminopolycarboxylic acids (EDTA, NTA, …) complexing agents or in frontal analysis mode (Kolodynska and Hubicki, 2012).

As indicated with the complexing agents mentioned above, chelating resins of IDA, AMP, phosphoric, phosphinic types, Diphonix® resins (Kolodynska and Hubicki, 2012) as well as D2EHPA containing resins (such as Lewatit® VO-OC 1026) have varying degrees of affinities for REE and can be used for their separation and purification. The difference in REE separation between cation exchange resins and complexing resins is that the affinity of the REE for conventional SAC exchange resins is about the same and separation is essentially based on the choice of the eluents. On the other hand, their affinities for complexing resins are different and a first separation takes place already during loading.

The recovery and separation of rare earths from phosphoric acid using phosphinic and phosphoric type resins have been studied (Kumar et al, 2010). It was found that extraction of rare earths from phosphoric acid decreases as phosphoric acid concentration increases and in addition, different metals showed different extraction at the same acidity, indicating separation possibilities. In fact, the percent extraction was found to decrease with increasing ionic radii of the RRE.

The use of IDA type resins in separation of REE presents the interesting aspect in that the affinity sequence of IDA resins is high affinity for light REE and low affinity for heavy REE while EDTA, which has similar structure as the functional groups of IDA resins has the opposite: high affinity for the heavy REE and low affinity for the light REE. This difference is used to separate heavy REE using IDA resins and EDTA as eluent (Moore, 2000).

Gallium

Gallium is obtained as a by-product from the Bayer liquor in the alumina production. There are both SX and IER processes for the recovery of Ga. The solvent used in SX is a derivative of 8-quinolinol, the 7-(4-ethyl-1-methyloctyl)-8-quinolinol, called Kelex 100. Ga is stripped from the solvent with HCl or H_2SO_4. In the ore (bauxite), the ratio Al/Ga is about 3000. It becomes

240 in the Bayer liquor and in the extract from the solvent it is about one.

There are two IER techniques to extract Ga from the Bayer liquor :
- a SIR impregnated with Kelex 100
- a chelating resin having amidoxime functional groups

Of these two resins, the Kelex-impregnated adsorbent has higher Ga-capacity and longer life time than the amidoxime resin, but its cost is significantly higher.

Acid Mine Drainage

Acid mine drainage (AMD) is acidic water coming off from abandoned mines or tailings piles. Sulfide minerals, after coming in contact with air and water generate acidity from the oxidation of the metal sulfides. Depending on the mineral, metal ions dissolved in the acidic waters may be iron, copper, zinc or nickel.

Pyrite oxidation can be described with the following reactions:
$$2\ FeS_2 + 7\ O_2 + 2\ H_2O \longrightarrow 2\ Fe^{2+} + 4\ SO_4^{2-} + 4\ H^+$$
$$4\ Fe^{2+} + O_2 + 4\ H^+ \longrightarrow 4\ Fe^{3+} + 2\ H_2O$$
The formed acidity keeps Fe^{3+} in solution.

There are various treatments for AMD, the most frequent being lime precipitation. According to this method, a slurry of lime is dispersed in a tank with AMD where pH increases to about 9. At this pH the metals are precipitated as hydroxides. The slurry is then directed to a clarifier where clean water will overflow for release while settled materials (sluge) come out at the bottom. The sludge contains metal precipitates, gypsum and unreacted lime. Part is recycled to the lime treatment tank and part is discarded.

Other, non-ion exchange, treatments include carbonate precipitation, metal sulfides precipitation and sulphate reducing biological treatment.

One disadvantage of precipitation methods is that they result in a water of high salinity and consequently of limited use. In one process, a combination of precipitation and ion exchange resins is employed. Heavy metals are first precipitated by raising the pH and then the salinity of the water is reduced by conventional SAC and WBA exchangers (Gaikwad and Gupta, 2008). A gel-type SAC resin and a high capacity WBA resin are used, as in water demineralization. To make the process economical, H_2SO_4 can be employed to regenerate the SAC and a mixture of NaOH and lime to regenerate the WBA resin. The generated acidic and alkaline gypsum from the SAC and WBA regeneration can be combined to produce good quality gypsum for further use.

An analogous process would be the use of SAC and WBA resins to treat directly AMD where the SAC removes all cations eventually preferentially the heavy metal cations due to a higher selectivity over other cations present, and the WBA removes preferentially any uranyl sulphates, sulphates and other anions. The spent regenerants contain then the toxic metals along with salts from the spent regenerants but in a concentrated form.

The use however of conventional SAC to remove heavy metals directly from AMD it may not be very easy because the selectivity differences between the heavy metals and other cations such as Ca^{2+}, Mg^{2+}, Na^+ etc is not that great and the heavy metals are found in lower concentrations in the AMD than other cations, together with high H^+ concentration.

The selectivity sequence for a conventional SAC resin is (de Dardel and Arden, 1989):

$H^+ < Na^+ < NH_4^+ < Mn^{2+} < K^+ < Mg^{2+} < Fe^{2+} < Zn^{2+} < Co^{2+} < Cu^{2+} < Cd^{2+} < Ni^{2+} < Ca^{2+} < Sr^{2+} < Hg^{2+} < Ag^+ < Pb^{2+} < Ba^{2+}$

Since Ca^{2+} is expected to be at a high concentration, it may displace metals like Mn^{2+}, Zn^{2+} etc, according to the above selectivity sequence, and there may be a pick of high concentration of metals, if the cycle does not stop in time.

A selective removal of heavy metals from AMD may be better achieved using selective resins.

Copper can be removed from the acidic pH of AMD effluents using an IDA type resin. Nickel, cobalt, zinc or cadmium may also be removed directly from many AMD solutions with IDA resins (at pH about 3-3.5). The presence of Fe^{3+} however is troublesome because it is removed preferentially than copper, nickel or cobalt. Selective removal of Fe^{3+} in presence of Cu^{2+} can be achieved with Diphonix® resins. Another way is to neutralize to a pH of 7, precipitate Fe^{3+} and extract Zn, Cd and Pb from the supernatant.

Uranium and thorium can be removed from AMD with SBA resins because these metals form stable anionic complexes with sulphates.

Zinc from geothermal brine

One intriguing application of ion exchange is the recovery of zinc from geothermal brine. These brines are sources of clean, low cost energy and a 330MW geothermal generating station has been operating at CalEnergy's Salton Sea location in the USA for a number of years. Because this brine contains significant quantities of zinc, CalEnergy has installed a process to harvest this zinc, prior to re-injection of the brine into the geothermal field. Using a combination of ion exchange, solvent extraction, and electrowinning, the plant has a design capacity for recovery of about 30,000 metric tons of zinc per year. It may also be possible to recover other dissolved metals such as manganese.

Due to the high chloride concentration of these geothermal brines, the zinc is typically present as the tetrachloro anionic complex ($ZnCl_4^{2-}$) which can therefore be effectively removed by a strong-base anion exchange resin in the chloride form. Elution of zinc is then readily accomplished by rinsing the resin with water which, because of the drop in the chloride concentration, breaks the anionic complex and allows Zn^{2+} to be released by the resin and to be recovered by electrowinning.

5. Purification of chemicals and process streams.

Brine purification in the chloralkali industry

The production of chlorine and caustic soda is based on the electrolysis of salt. There exist three process technologies for the electrolytic production of caustic and chlorine from salt. These are the diaphragm cell, the mercury cell and the membrane cell process. In all cases, on the anode, Cl^- ions discharge to give Cl_2

$$2\ Cl^- \longrightarrow Cl_2 + 2\ e^-$$

On the cathode, in the cases of diaphragm and membrane cells, water gives OH^- and hydrogen :

$$2\ H_2O + 2\ e^- \longrightarrow 2\ OH^- + H_2$$

The diaphragm or the membrane serve to separate anode and cathode to prevent Cl_2 to react with H_2 or with NaOH. In mercury cell, NaOH is produced separately: the Na^+ ions on the mercury electrode form an amalgam with the mercury which then decomposes in water to give H_2 and OH^- :

$$Na^+ + e^- + Hg \longrightarrow Na(Hg)$$
$$2\ Na(Hg) + 2\ H_2O \longrightarrow 2\ OH^- + H_2 + 2Na^+ + 2(Hg)$$

The diaphragm (asbestos) is not selectively permeable so the produced NaOH is contaminated with NaCl. The purification of NaOH is achieved by evaporation to 50% NaOH where NaCl precipitates out. Still, the NaOH contains some 1% NaCl. Removal of Cl⁻ from concentrated NaOH can be done with a process based on ion retardation (page 218). In membrane cells, the produced NaOH is 33% and practically NaCl-free. This NaOH is subsequently evaporated to 50% NaOH. Purification of NaOH from last traces (about 20 ppm) of NaCl can be done with the ion retardation technique, as discussed later.

Today, the production of Cl_2 and NaOH is based on the diaphragm and membrane cells technologies. The mercury-cell technology has been phased out.

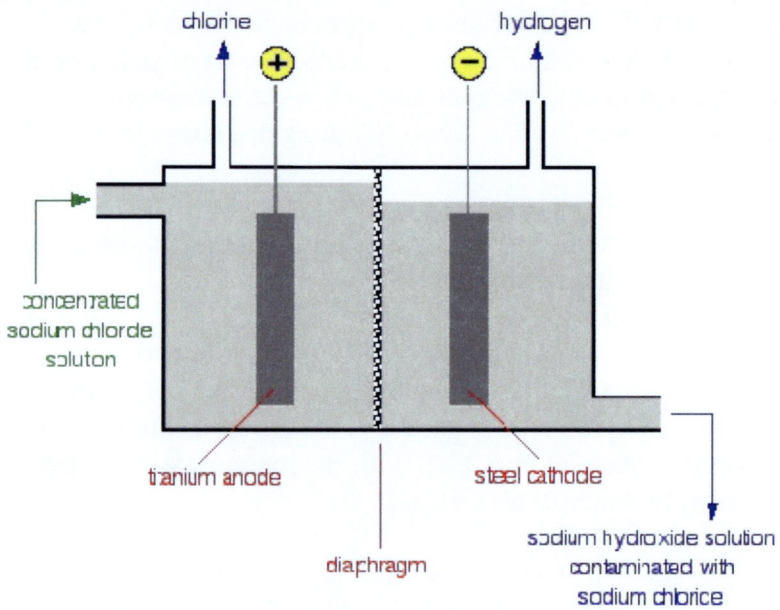

Figure 5.1 Diaphragm cell

One side reaction at the anode compartment is the formation of chlorates:

$$Cl_2 + 2\ NaOH \rightarrow NaOCl + NaCl + H_2O$$
$$3\ NaOCl \rightarrow NaClO_3 + 2\ NaCl$$

In membrane cells, since the depleted NaCl is recycled, there is a built-up of $NaClO_3$ which eventually causes membrane inefficiencies and requires purging.
One way to remove chlorates from brine is (Moore and Dotson, 1984) to circulate part of the depleted brine, after dechlorination and saturation with NaCl, and adding HCl whereby $NaClO_3$ is decomposed to Cl_2 and NaCl:

$$NaClO_3 + 6\ HCl \rightarrow NaCl + 3\ Cl_2 + 3\ H_2O$$

At lower ctoichiometric quantities of HCl, ClO_2 is formed:
$$NaClO_3 + 2\ HCl \rightarrow ClO_2 + ½\ Cl_2 + H_2O + NaCl$$

Regardless of the type of cell employed, a suitable brine must be prepared prior to entering the electrolysis cells because impurities in the brine can affect the performance of all types of cells. Membranes however are especially sensitive to brine impurities and the membrane cell process requires a higher degree of brine purification than any of the other technologies.
In a membrane cell (Fig. 5.2), a cation exchange membrane (CEM) is used as a separator of the anode and the cathode. The CEM allows the passage of cations, such as Na^+, but not of the anions, such as Cl^- or OH^-. Impurities can cause current efficiency decline by reducing the resistance to allow the anions to pass through the CEM. This decrease in anion rejection can be

caused by membrane damage, which is caused by precipitation of impurities in the membrane (Keating and Behling, 1990).

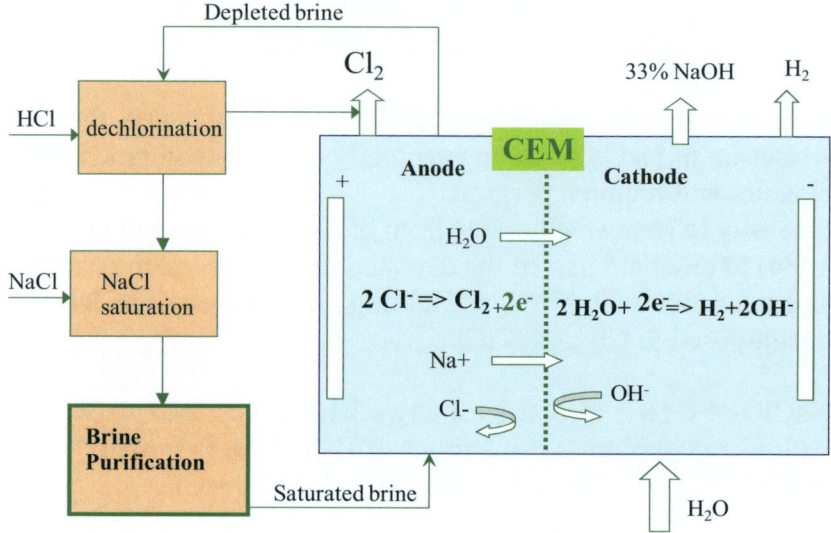

Figure 5.2   membrane cell technology

There are different types of salt used as raw material for the brine. It can be underground rock salt, solar salt, obtained by evaporation of saline or sea waters, or vacuum purified salt, obtained by crystallization of salt from brines. These different salts contain different impurity levels. The main impurities encountered are calcium, magnesium, strontium, barium, iron, aluminum, sulfates, silica, trace metals, iodides. Aluminum alone is not harmful, however in combination with silica can form aluminosilicates inside the membrane. Similarly, silica sensitizes the membrane to calcium by forming calcium silicate in the membrane. Iodides can form the highly insoluble sodium paraperiodate, $Na_3H_2IO_6$, or interact with barium. In the case where mercury cells have been converted to membrane cells and for a certain period of time both types of cells co-exist and share the same brine, mercury can also be another impurity.

After dissolving the solid salt in the salt dissolver, the first step in brine purification is the addition of sodium carbonate and caustic soda to precipitate calcium carbonate, magnesium and iron hydroxides. These calcium carbonate precipitates entrain also the hydroxides of aluminum, strontium and magnesium. These precipitates are settled in a settler and the overflow, containing some 10-50 ppm suspended solids, is filtered.

Table 1 gives a typical composition of the filtered brine after the primary purification.

TABLE 1 Typical brine composition after primary purification

Component	concentration
NaCl	300-315 g/L
Ca	<5 ppm
Mg	<1 ppm
Sr	<3 ppm
Ba	<0.5 ppm
Fe	<0.1 ppm

In the case where vacuum salt is used as raw material, the precipitation step is not necessary due to the higher purity of this salt. In some cases, vacuum salt contains sodium ferrocyanide as anticaking agent. In this case, the iron cyanide complex can be oxidized in the anode compartment by Cl_2 releasing iron. Subsequently, iron can pass to the cathode compartment where it can precipitate as hydroxide. In order to remove the iron, the cyanide complex is oxidized before the primary purification and the resulting iron hydroxide is removed by filtration.

Even though this filtered brine is suitable for feeding mercury and diaphragm cells, for membrane cells a higher purity is required, for example and depending on the membrane type:

Ca, Mg <20 ppb
Sr < 50-200 ppb
Ba< 100-500 ppb
Iodine < 0.2 ppm
Nickel < 50 ppb

In order to obtain a brine with the impurity levels indicated above, ion exchange resins are employed (White and O'Brien, 1990).

Hardness removal

Removing of hardness ions, calcium, magnesium, strontium and barium, from brine, necessitates the use of resins having extremely high selectivities for the hardness ions with respect to the sodium ions. Conventional strong acid cation exchangers used in water softening would not be appropriate here due to the very high Na^+ concentration in the brine. Two types of resins containing ligands as functional groups are used commercially for brine softening. One type has iminodiacetic (IDA) functional groups and the other aminomethylphosphonic (AMP) functional groups. Both types come in a macroreticular (macroporous) structure which gives to these resins an easier accessibility of the functional groups and a better physical stability.

Both these groups can form very stable complexes with calcium, magnesium, strontium and barium at pH values 9-11 and therefore these hardness elements can be fixed on the resin from brine solutions containing a few ppm hardness (fraction of a meq/L) even in the presence of 5-6000 meq/L of Na^+.
The ion exchange resins are positioned after the primary purification (figure 5.3):

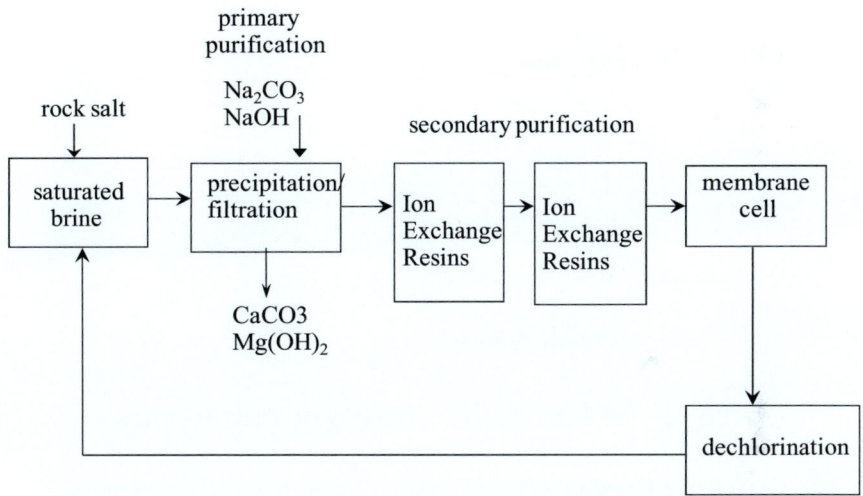

Figure 5.3 Secondary brine purification: hardness removal

The system used is either two columns in series in lead-lag configuration, or three columns in series in merry-go-round configuration. In the first case, when the lead column breaks through, it goes to regeneration while the lag column goes to the lead position. After the regeneration of the first column is finished, the column goes to the lag position. In the merry-go-round, there are always two columns in service and one on regeneration or stand by (figure 5.4):

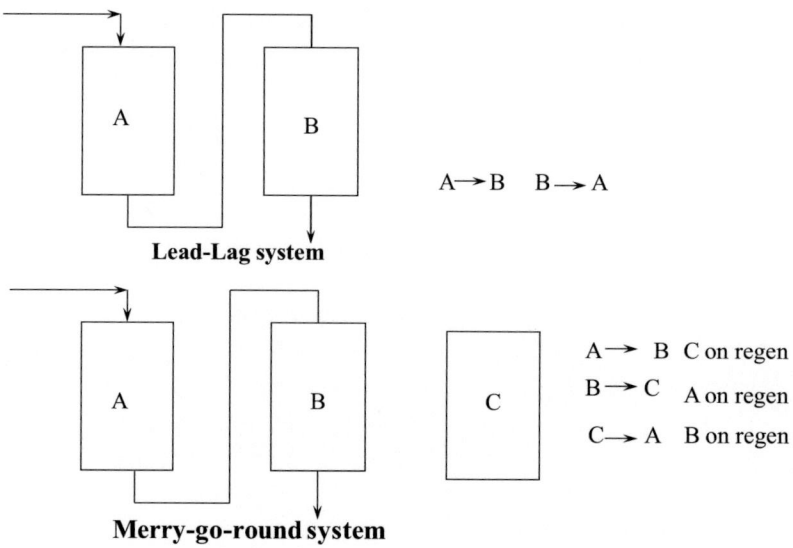

Figure 5.4 Lead-lag or merry-go-round systems

The removal of hardness ions from concentrated brines is in general a slow process. Parameters such as resin particle size and porosity affect the kinetics of hardness removal and have an effect on leakage and capacity. In order to improve kinetics and physical stability, certain resin manufacturers have introduced IDA and AMP selective resins having a uniform particle size of 0.39 mm (Lanxess, 2013).

Two fundamental differences between these two types of resins are, as mentioned in page 45:

The IDA type resins have higher affinity for Sr^{2+} and Ba^{2+} compared to the AMP type resins

The AMP type resins have higher affinities for Ca^{2+} and Mg^{2+} compared to IDA type resins

Therefore, the choice of the resin depends on the feed brine composition and the specifications for the treated brine. If the brine contains high Sr levels, IDA type resin is preferred, otherwise AMP type resin can be used. For both types however, the relative selectivities for the hardness ions under the typical operating conditions for pH and temperature are:

$$Mg^{2+}>Ca^{2+}>Sr^{2+}>Ba^{2+}$$

Figure 5.5 illustrates qualitatively the difference in selectivities for Sr^{2+} and Ca^{2+} in concentrated brines between AMP and IDA type resins:

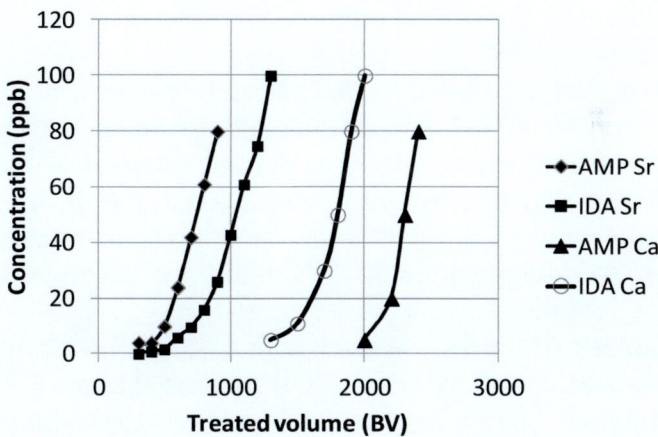

Figure 5.5 Brine softening with AMP and IDA type resins

There exist some additional factors that affect the resin choice. As mentioned above (p.183) vacuum or solar salts may contain sodium ferrocyanide, added as an anticaking agent. The ferrocyanide being an anion, is not fixed on the chelating resins. In the anolyte, the cyanides are oxidized and release the iron as a cation. At the same time, the nitrogen forms N-containing compounds which are very unstable and constitute a risk for the installation. If there is no oxidation and precipitation to oxidize the ferrocyanide and remove Fe^{3+} as $Fe(OH)_3$, the iron can pass through the chelating resins where it can be fixed. In the case of AMP type of resin, the iron is fixed irreversibly while in the IDA types iron can be eluted during the HCl regeneration step.

Another element that can poison the AMP type resins is mercury. The removal of mercury from the brine and its effect on the hardness-removing resins is discussed later. Therefore, in cases where iron or mercury are found in the brine going to the chelating resins, IDA type resins present a longer life time.

Important design parameters that affect the performance of the ion exchangers are pH, temperature and flow rate. The operating capacity of the resins increases as pH and temperature increase. Similarly, the operating capacity decreases as the specific flow rate, expressed as BV/h, increases. An optimal set of conditions is pH of 9-11, temperature 60-70°C and flow rate 20-30 BV/h.

Regeneration of the resin is done with 4-8% HCl, whereby the resin releases all cations fixed and is converted to the H^+ form. Following this, the HCl is rinsed with softened water and the resin is converted to the Na^+ form using 4% NaOH. Prior to regeneration, the brine is displaced from the resin using softened water. This step is important because when the HCl enters the column, the chloarate ions that may contain the remaining brine in the column, will react with the acid resulting in a release of free chlorine which will oxidize the resin and thus shorten its life time.

Mercury removal

In cases where chlor-alkali plants combine mercury and membrane technologies, mercury can be a contaminant for the feed brine to the ion exchange resins of the secondary purification. The effect of mercury on the ion exchangers has been studied (Yamaguchi, 1990). It was found that both types of resins, IDA and AMP types, load mercury. During normal regeneration, that is with 6-8% HCl followed by the NaOH conditioning, the amount of mercury regenerated from the IDA type resins is significantly higher than that regenerated from AMP type resins. With IDA type resins, a steady state is reached rather quickly (about 10 cycles) where the amount of mercury regenerated equals to the amount loaded during service cycle and the capacity taken by the irreversibly fixed mercury represents about 25% of the total capacity. The AMP type however, the amount of eluted mercury at every regeneration is small so that the cumulative mercury loading on the resin results in a resin fouling. It is conceivable therefore that an IDA type resin can be used to remove mercury as well as hardness, provided that the installation is designed to contain about 25% more resin.

Another option is to use an ion exchanger that selectively removes mercury from brine. Such resins are the those described in pages 52-56 having sulfur-containing ligands as functional groups.

The resin fixes only ionic mercury, which is the form usually found in the brine. The design of such a unit is based on a flow rate during the loading step of 10 BV/h. The system used is two columns in series in lead-lag configuration and is placed before the softening resins. If the resin is regenerable, like thiol resins, then when one column is on regeneration, the other assures the removal of mercury. With non-regenerable resins (thiourea or thiouronium types) then the exhausted resin is replaced with new one. With a Hg^{2+} concentration of 10 ppm in the feed solution, at a specific flow rate of 10 BV/h, the cycle length can be about one month or two depending on the resin.

The presence of two columns in series with the second column as a polisher, assures a final leakage of less than 5 ppb.

Regeneration of thiol resins is performed with concentrated (30-35%) HCl.

Aluminum and silica removal

Aluminum and silica are found in brines at levels of 0.1 to 2.5 ppm and 0.1 to 20 ppm respectively. The removal of aluminum and silica from brines can be accomplished with chelating resins such as IDA or AMP types (Kelly, 1984). The brine is first acidified to a pH of 2 to 3 in order to break any nonionic colloidal complexes of aluminosilicates and solubilize Al^{3+} and SiO_2. Then it passes through the resins in the Na form at a flow rate of 10 BV/h. The resin can be placed after the primary purification.

Due to the fact that the resin is found in the Na^+ form, the effluent pH is in the range of 6, where Al^{3+} is well fixed by the resin and well regenerated subsequently. SiO_2 is removed at the same time at a level of about 50% of the feed concentration while the Al^{3+} leakage is less than 0.1 ppm. Breakthrough of Al^{3+} takes place when the effluent pH drops below about 5.5 It should be noted that if Al^{3+} is fixed while the pH attains low values (around 2), the elution during the regeneration step is less efficient. The loading capacity for Al^{3+} is between 2 and 6 g/Liter resin.

Regeneration and Na conversion of the resin is performed using 4-8% HCl and 4% NaOH, in a similar way as in the brine softening operation

Nickel removal

Nickel in brine fouls the membranes and must be removed down to ppb levels. The removal of nickel from brine is achieved with an IDA or AMP type chelating resins after the softening step by adjusting the pH to 3 and passing through the resin at a specific flow rate of 20 BV/h and temperature 60-80°C. From a nickel concentration of around 30 ppb, after the resins it goes to < 10 ppb. Due to the very long cycle, the resin can be used only once, with no regeneration.

Iodide removal

Iodides are present in brines and can precipitate in the form of the highly insoluble sodium paraperiodate,

Na$_3$H$_2$IO$_6$ causing loss in current efficiency in the electrolysis cells. Membrane suppliers require less than 0.2 ppm of I$^-$ in the feed brine of membrane electrolysers. The principle of the process for removing I$^-$ from brines is a patented process according to which I$^-$ are first oxidized to I$_2$ using chlorine and then fix the (I$_2$Cl)$^-$ thus formed on a strong base anion exchange resin in the Cl$^-$ form (Arrighi and Pastacaldi, 1998)

$$2\,I^- + Cl_2 \rightarrow I_2 + 2Cl^-$$
$$I_2 + Cl^- \rightarrow (I_2Cl)^-$$
$$R^+Cl^- + (I_2Cl)^- \rightarrow R^+(I_2Cl)^- + Cl^-$$

The brine is first purified from calcium and magnesium. Then it is acidified to a pH<3, usually around pH=2, and oxidized using an in-line electrolyser. The redox potential is controlled at 550-600 mV in order to avoid the formation of the unwanted iodates (Figure 5.6).

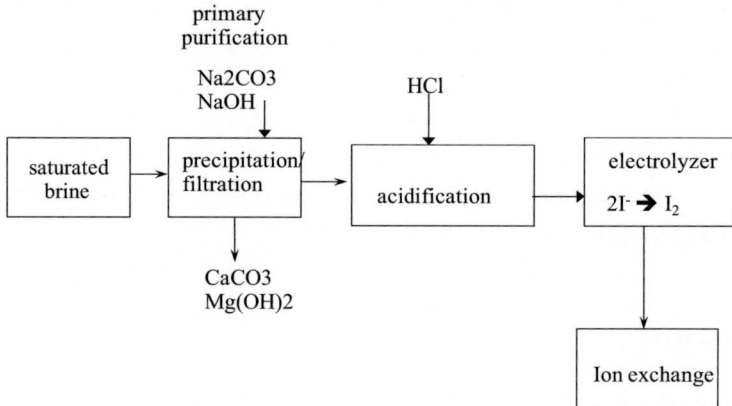

Figure 5.6 Iodides removal from brine

The loading step is performed at a flow rate of 2-4 BV/h. The ion exchange resin can be a gel type or a macroreticular type. The useful capacity that can be obtained depends upon the breakthrough end-point, the loading flow rate and the regeneration procedure, however, loading capacities can be very high probably because of the formation of polyiodine chains fixed on the resin.

Regeneration can be done with 2-4 BV of a 12% Na_2SO_3 solution that reduces the I_2 to I^- which is then eluted from the resin. Other regeneration procedures have been suggested (Nakamura *et al* 1964) where NaOH is first applied to elute I_2 followed by NaCl which removes the rest of I^- remained on the resin.

$$3 I_2 + 6\ NaOH \rightarrow 5\ NaI + NaIO_3 + 3\ H_2O$$

Iodine is then removed from the spent NaCl solution by acidification with an acid, HX, and crystallization of I_2.

$$5\ NaI + NaIO_3 + HX \rightarrow 3\ I_2 + 6\ NaX + 3\ H_2O$$

Sulfates removal

The presence of sulfates causes deposits in the membrane in the cathode side and reduces the current efficiency.
Common methods to control the sulfate concentration in the brine is by purging the brine system or by adding $BaCl_2$ which precipitates sulfates.

The use of nanofiltration membranes in sulfate removal from brines is another feasible method (ten Kate et al, 2009).
Another method is to use a SAC resin loaded with hydrous zirconium hydroxide (Lee and Bauman, 1983).
The use of conventional anion exchange resins are not selective enough to remove SO_4^{2-} from brines, either the saturated (about 300 g/L NaCl) or the depleted brine (200 g/L NaCl), because of a too high concentration of the Cl^-. In a patent (Minz and Vajna, 1985) the depleted brine is diluted with a water quantity equivalent to the water losses during the electrolysis process. This diluted brine, having a NaCl concentration of about 100 g/L is then treated with a WBA exchange resin at a pH of 3 to remove sulfates. The resin is regenerated with 300 g/L NaCl brine and the Na_2SO_4 is crystallized off in a cooling crystallizer so that the brine regenerant is recycled.

Nippon Rensui Co. and Mitsubishi Chemical Corp. have developed a chromatographic separation method using the amphoteric ion exchange resin Diaion® DSR01 (Matsushita, 1996). This resin contain strong base and weak acid groups on the same quaternary ammonium groups, and not interpenetrating networks of strong base and weak acid resins, as described earlier (Dowex® Retardion® 11A8, page 74):

They have shown the following selectivity sequence:

Chlorate> chloride> sulfate

As a consequence, sulfates are eluted first, followed by chlorides and then by chlorates.

Lithium brines

Lithium is an alkali metal with industrial applications including ceramics, alloys, lithium batteries and lithium-ion batteries. Li is produced electrolyticaly from fused LiCl/KCl salts at 450°C. Lithium is found in Li containing brines from which it can be recovered in pure form. Pure Li_2CO_3 is prepared from brines after precipitating Mg with KCl as carnallite, then adding Na_2CO_3 to precipitate Ca, filtering, precipitating Li_2CO_3 by evaporation, redissolve, add CO_2 to form a solution of $LiHCO_3$ which is purified through a low sodium IDA or AMP type resin to remove traces of Ca and Mg (Holger *et al*, 2003; Boryta and Donaldson, 2012). Then form Li_2CO_3 by heating, which is recovered by evaporation and crystallization.

In an interesting patent (Rezkallah, 2015) separation of lithium and magnesium was achieved with elution chromatography using a chromatographic grade low DVB SAC exchange resin. According to this patent, magnesium and lithium are separated by water elution where magnesium comes out ahead of lithium.

In addition to the above, a boron-selective resin can be used to remove boron from the $LiHCO_3$ liquor (Harrison and Blanchet, 2011).

LiCl is prepared from purified Li_2CO_3 by reacting with HCl.

KCl brines

KCl, the raw material for KOH production by electrolysis is purified in a similar way as NaCl brines. However, the resin characteristics in KCl brines purification are not necessarily the same as in NaCl brines. For example, the swelling of the chelating resins of both, IDA and AMP types is higher going from the H^+ to the K^+ form than going from the H^+ to the Na^+ form. This can result in a shorter resin life time due to breakage of the beads.

Boron removal from $MgCl_2$ brines

One of the processes to produce magnesia (MgO) is by hydropyrolysis of magnesium chloride:

$$MgCl_2 + H_2O \xrightarrow{450\text{-}1000°C} MgO + 2\,HCl$$

In order to form refractory magnesia, boron has to be removed from $MgCl_2$ brines to avoid embrittlement of magnesia tiles. Similarly, boron has to be removed from $MgCl_2$ brines for the electrolysis of fused $MgCl_2$ for the production of magnesium metal. For both cases, a boron-selective resin (page 59) can be employed. Also, a boron-selective resin can be used to remove boron from the $LiHCO_3$ liquors (Harrison and Blanchet, 2011) for the production of pure LiCl.

Boron removal from solutions is in general a slow process and the kinetics are particle diffusion controlled. The water contained in a resin in equilibrium with pure water is called moisture holding capacity (MHC), or water uptake or other similar term, and is one of the principal characteristic properties of the resin. The MHC of a resin depends upon the crosslinking density of the polymer matrix, the nature of the polymer, the nature and the concentration of the ionic functional groups (the total exchange capacity) and the nature of the counter-ions. When an IER comes in contact with an aqueous solution rather than with pure water, the resin swells less than in pure water because the osmotic pressure difference between the resin and the external solution is smaller. If the external solution is very concentrated, as is the case here where concentrated $MgCl_2$ brines are treated, the resin shrinks as water is displaced from the resin to the external solution. In these cases the "free" water inside the resin beads, or inside the gel microbeads of a macroreticular type resin, is considerably reduced and the kinetics of boron removal is reduced as well due to slower diffusion. It is conceivable also that the resin in the $MgCl_2$ brine gives a different (lower) equilibrium capacity compared to the equilibrium capacity in water. The performance (meaning essentially operating capacity to a given break-through concentration and boron leakage) of the boron-selective resins is controlled to a large extent by the specific flow rate (BV/h), by the brine concentration (which controls the "free" water inside the resin beads) and by the particle size of the resin, in addition of course to the regeneration procedure and regenerant levels. At $MgCl_2$ concentrations approaching the saturation values (about 460 g $MgCl_2$/L at ambient temperature) where in addition the viscosity of the solution starts increasing, the dehydration of the resin is such that the operating capacity in boron removal becomes close to zero.

The regeneration of the boron selective resins is performed in two stages. First, 10% H_2SO_4 is used to dissociate the boron complex and to elute boron from the resin. A level of 50 g H_2SO_4/L_R is enough to regenerate the resin. After the acid step, the resin is found in the HSO_4^-/SO_4^{2-} form and it gives an acidic pH. In order to convert the resin to the free base form, a caustic wash is used at a level of 36 g $NaOH/L_R$ using a 4% NaOH solution. If there is no caustic wash, the boron leakage will be higher due to the acidic pH of the resin and the operating capacity will be lower by about 15%. Instead of H_2SO_4, HCl can be used at a level of 35-40 gHCl/L_R using a 4% HCl solution.

Figure 5.7 illustrates the effect of $MgCl_2$ concentration on the operating capacity of a boron-selective resin.

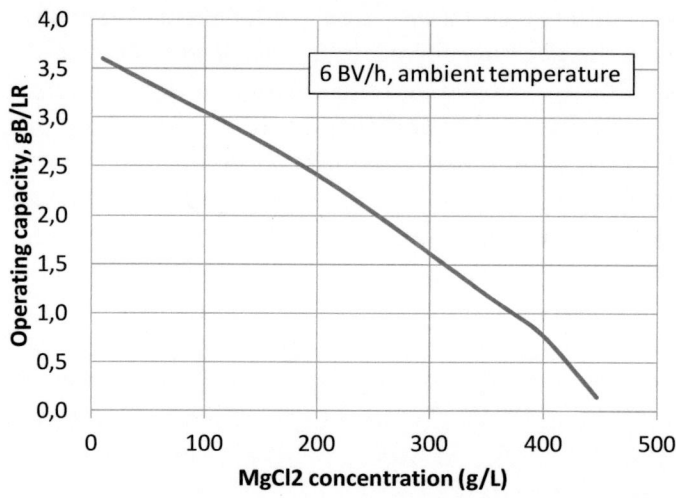

Figure 5.7 Effect of $MgCl_2$ concentration on capacity of boron-selective resins

The curve gives only an indication of the effect since parameters such as regeneration type and level which are very important are not defined. Also, different commercial products having different characteristics such as MHC, particle size etc, may give different values for operating capacity.

Figure 5.8 gives the effect of the specific flow rate on the operating capacity of a boron-selective resin, for both, water and concentrated brine solutions. Again, as explained above, this figure only illustrates this effect.

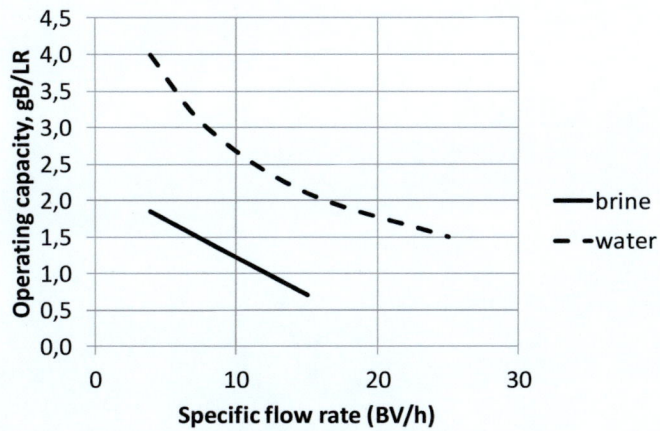

Figure 5.8 Effect of specific flow rate on capacity of boron-selective resins

Hydrogen Peroxide purification

H_2O_2 is prepared via the anthraquinone process as shown below. After the autoxidation process, anthraquinone is formed and H_2O_2 is recovered from the reaction mixture with water and distilled to obtain industrial pure grade (IPG). This grade is sufficient for food and pharmaceutical applications. For other applications such as electronics industry where high purity is required (<100 ppt of cations), or paper industries, the H_2O_2 is further purified by another distillation followed by organics removal with an adsorbent followed eventually by a demineralization unit to remove cation and anion impurities.

IPG H_2O_2 contains as contaminants Na, Ca, Zn, Fe, Al, transition metals and organics (anthraquinone derivatives). For the removal of organics, a styrene/DVB adsorbent is used (Honig and Geigel, 1995). Due to the strong oxidizing properties of H_2O_2, the conditions of use of the organic adsorbent as well as the demineralization unit must be carefully chosen and tested beforehand. It should be remembered that the natural pH of H_2O_2 is around 4. At pH of 6-7 the decomposition rate of H_2O_2

increases considerably. At a pH of 10 H_2O_2 exploses! Transition metals catalyze the decomposition of H_2O_2.

The conditions of use in organics removal with a synthetic adsorbent are summarized as follows: the adsorbent to use should have relatively small pores, enough however to allow anthraquinone derivatives to enter easily. The H_2O_2 concentration that comes in contact with the adsorbent should be below 35%. Flow rate is 3-4 BV/h. The operating capacity with a typical commercial adsorbent is 30 g C/L_R with a feed concentration of about 300 mg C/L (here the concentration is expressed as carbon, C) to a leakage of 20% of the feed concentration. The adsorbent usually swells in H_2O_2 and therefore it is important to know the degree of swelling with the specific adsorbent used to take into account in the design of the installation and for operational steps (backwash). The regeneration is performed with 2-4 BV of methanol followed by a long rinse with water (20-30 BV) to bring down the methanol concentration to very low levels. Great care must be exercised during resin installation to rinse away the salts and other preservatives from the fresh adsorbent resin because some of the salts may give an alkaline pH. This requires an extremely long water rinse until no more salts come out of the adsorbent. Similarly, special pretreatment is required if high quality H_2O_2 is produced (Hoffman, 2003).
After a certain time of use, the adsorbent builts up carboxylic groups as a result of oxidation by the H_2O_2. Any metal impurities can be fixed on these carboxylic groups which built up since they are not regenerated with the methanol. These metals can catalyze further oxidation of the adsorbent. Occasionally, slightly acidified methanol can be used as regenerant to remove the accumulated metals. Typical life time of the resin is considered as 2-3 years, or more if low operational temperatures are used.

For H_2O_2 used in the electronics industry, like washing silicon wafers, a highly pure quality is required (sub ppb levels). According to one technique (Sakaitani et al, 1997) the hydrogen peroxide is treated with a chelating resin of AMP type to remove iron and aluminum ions from a level of several ppb down to 0.1 ppb. If phosphates are present above 0.1 ppm then a SBA resin precedes the chelating resin. The SBA resin is in the HCO_3^- form (never in OH^- form because H_2O_2 explodes at high pH !!!) and it removes phosphate anions down to 0.05 ppm.

In order to deionize hydrogen peroxide using conventional ion exchange resins, the operating conditions are chosen so to eliminate any accident from the contact of H_2O_2 with the ion exchangers during operation. In a patent to Interox Chemicals Limited (Millar et al, 1995) the deionization is performed with ion exchangers by limiting the contact time of H_2O_2 with the resins, that is, by passing the solution at specific flow rates > 200 BV/h and even at 500-2500 BV/h, by using equipment vented to the atmosphere and by slurrying the resin bed. The H_2O_2 solution is allowed to pass many times through the resin bed but the resin is replaced when the operating capacity reaches the 20% of the total capacity.

After the SAC, a SBA in the HCO_3^- form can be used to remove the acids formed during the passage from the SAC resin. The H_2O_2 concentration is less than 30-40%.

In another patent by Air Liquide (Devos et al, 2001) H_2O_2 passes first through the SBA in HCO_3^- form in upflow direction, in order to facilitate the removal of CO_2 gas that may be formed during the exchange of the anions with the HCO_3^- anions. It then passes through the SAC resin in the H^+ form. Temperature is 5°C, flow rate 15 BV/h and H_2O_2 concentration 30%.

Dimethyl formamide purification

Dimethyl formamide (DMF) is a good solvent for polyacrylonitrile (PAN). In making PAN filaments, PAN dissolved in anhydrous DMF is extruded, or dry-spun. DMF is recovered as an aqueous solution and is concentrated by distillation. Going through the extrusion and the recovery phases, DMF is found at high temperatures and impurities are formed through hydrolysis or other reactions. In order to recover DMF and recycle it back to process, the anhydrous DMF can be treated with a SAC resin followed by a WBA (Hewett, 1960). In general, MR type resins are used. After the loading cycle, it follows a sweetening off step with water, then regenerant injections, rinse and then sweetening on with DMF. The sweetening on and off liquors can be recycled to the DMF distillation.

A WBA resin and not a SBA resin is used here because DMF hydrolyzes at high pH to give formic acid and dimethylamine. For that reason, after the caustic regeneration of the WBA resin, a NaCl wash can follow the rinse step in order to convert the strong base groups of the WBA resin to the Cl^- form without affecting the weak base groups, thus lowering the pH inside the resin beads.

As is the case when solvents come in contact with IER or adsorbents, the resin or adsorbent swelling in these solvents should first be determined before any column testing.

Glycols purification

Ethylene glycol (EG) is produced from ethylene in two stages. In the first stage, ethylene is oxidized directly with air or oxygen at high pressures and temperatures and in the presence of a heterogeneous catalyst frequently containing silver to form ethylene oxide (EO).
The produced ethylene oxide is removed from the reactor in a gas stream into an absorber where it is absorbed in water. The ethylene oxide containing water passes through a stripping column where it is stripped off with steam or hot water and goes to the ethylene glycol plant. The aqueous stripper bottoms are recycled to the absorber. The excess water is purged and since it contains 2-3% glycol it goes to the ethylene glycol plant.
In the second stage, ethylene oxide is hydrated at high temperatures to form ethylene glycol and this reaction can be catalyzed with acids or bases. The major by-products are diethylene glycol (DEG) and triethylene glycol (TEG). The monoethylene glycol (MEG) is separated from the higher glycols by distillation.

The production of MEG is illustrated in figure 5.9. Ion exchange can be used in the EO or EG plants in the following streams:

1. The purge (waste) water from the stripper in the EO plant (Broz, 1973). This water contains salts and 2-3% EG which can then be recovered. The ion exchange system to use is a conventional demineralization one: SAC exchanger followed by a degasser followed by a WBA exchange resin. Here macroreticular resins are preferred. As a polisher, a column of activated carbon can be used.

2. Removal of aldehydes from aqueous solutions of EG. This is achieved with SBA exchange resins in the bisulfite form (Mansoor, 2007). This approach is based on the reaction between bisulfite ions and aldehydes:

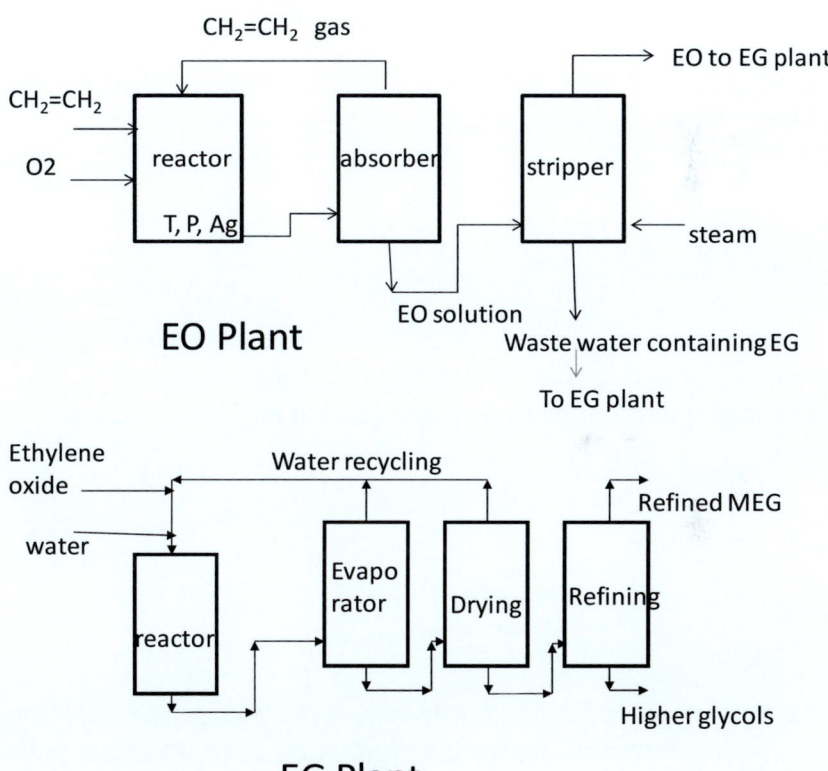

Figure 5.9 Ethylene glycol production flow sheet

$$\begin{array}{c} H \\ \diagdown \\ C{=}O \\ \diagup \\ H \end{array} + Na^+ HSO_3^- \longrightarrow \begin{array}{c} H OH \\ \diagdown\diagup \\ C \\ \diagup\diagdown \\ H SO_3^- Na^+ \end{array}$$

Thus, the reaction with the SBA resin in the HSO_3^- form is:

$$R^+ HSO_3^- + HCHO \longrightarrow R^+ HOCH_2SO_3^-$$

A gel type SBA resin can be used here at a specific flow rate of about 5 BV/h. Regeneration is done with 5% solution of $Na^+HSO_3^-$

$$R^+HOCH_2SO_3^- + Na^+HSO_3^- \longrightarrow R^+HSO_3^- + Na^+HOCH_2SO_3^-$$

3. Deacidification of the recycled water in the EG plant. Contains organic acids such as formic or acetic acids. A WBA exchange resin is used in this case (Rohm and Haas Internet site).

Caprolactam purification

The main application of caprolactam is the production of Nylon-6. The key intermediate for the production of caprolactam is the cyclohexanone oxime.

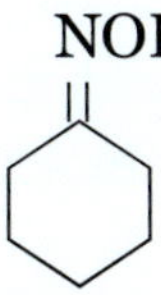

$$\underset{\text{Cyclohanone oxime}}{\text{[cyclohexanone oxime structure with =NOH]}}$$

Cyclohanone oxime

This can be obtained by the reaction of cyclohexanone with hydroxylamine to produce cyclohexanone oxime:

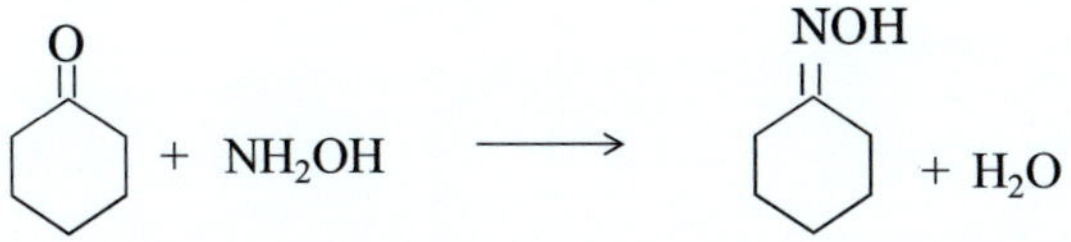

[cyclohexanone] + NH_2OH ⟶ [cyclohexanone oxime] + H_2O

Another route goes directly from cyclohexane to the oxime by photonitrozation of cyclohexane (PNC process):

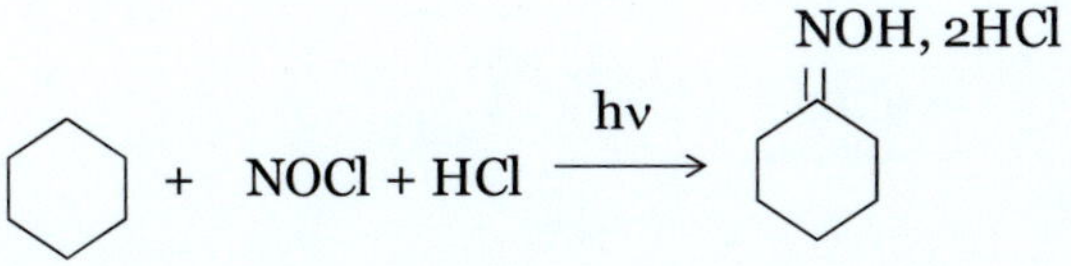

[cyclohexane] + NOCl + HCl $\xrightarrow{h\nu}$ [cyclohexanone oxime · 2HCl]

The cyclohexanone oxime undergoes a Beckmann rearrangement in the presence of oleum to produce caprolactam:

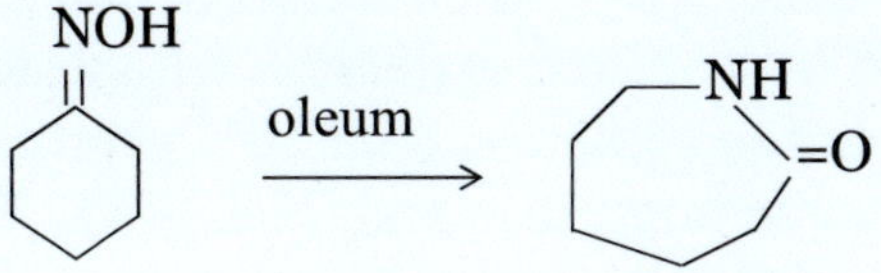

[cyclohexanone oxime] $\xrightarrow{\text{oleum}}$ [caprolactam]

caprolactam

The product of the Beckman rearrangement is the sulfate salt of caprolacam. This salt along with the excess of H_2SO_4 are neutralized with ammonia to produce free caprolactam and $(NH_4)_2SO_4$.

In another process developed by Sumitomo, cyclohexanone oxime is produced by ammoximation of cyclohexanone (EniChem process):

$$\text{cyclohexanone} + NH_3 + H_2O_2 \longrightarrow \text{cyclohexanone oxime} + 2\,H_2O$$

converted to caprolactam via vapor phase Beckman rearrangement using zeolite catalyst (Ichihashi and Sato, 2001):

$$\text{cyclohexanone oxime} \xrightarrow{\text{zeolite}} \text{caprolactam}$$

Therefore, in this process, the formation of $(NH_4)_2SO_4$ is avoided.

The crude caprolactam is purified by extraction with an organic solvent followed by a second extraction with water. The recovered aqueous solution contains about 30% caprolactam but it still contains organic impurities and $(NH_4)_2SO_4$. This solution is treated on ion exchange resins using a sequence SBA → SAC resins or SBA → SAC → SBA resins at a specific flow rate of 6-7 BV/h.

The regeneration procedure of the resins is specific to this process: the regeneration of each of the SBA starts with 5% HNO_3, rinse, then 4% NaOH, rinse. The regeneration of the SAC starts with 4% NaOH, rinse, then 5% HNO_3 (or 5% H_2SO_4 if there is no hardness) and rinse. The reason for this two-stage regenerations is to displace caprolactam from the resins before applying the suitable regenerant.

In another process where it is produced a mixture of caprolactam and laurolactam (Inaba *et al*, 1993; Hirosawa *et al*, 1995), after the neutralization with ammonia of the Beckman rearrangement mixture of the lactams, the two lactams are separated with liquid-liquid extraction where the caprolactam is recovered in an aqueous phase and the laurolactam in an organic phase. The caprolactam is subsequently purified with a SAC➔WBA➔SBA system at 6 BV/h specific flow rate. The impurities to be removed include, in addition, some alkylsulfonates used in the process. As WBA resin an acrylic type was employed, probably to be able to regenerate the resin more efficiently from the alkylsufonates. The other two resins were macroreticular styrene/DVB types. As regenerants, H_2SO_4 and NaOH were suggested, with or without the two-stage regeneration mentioned above.

Formaldehyde purification

Formaldehyde is produced from the catalytic oxidation of methanol at high temperatures:

$$CH_3OH \longrightarrow HCHO + H_2$$

Formic acid is formed as byproduct. The composition of the reaction mixture is 57-58% formaldehyde, 0.02-0.5% HCOOH and 1-20% methanol. Temperature 50°C. Formic acid can be removed with a WBA resin However if this resin contains some strong base groups, as is the case with many stryrenic type WBA resins, then if the strong base groups are in the OH⁻ form, then they can cause the Cannizzaro's reaction:

$$2\ HCHO + OH^- \longrightarrow HCOO^- + CH_3OH$$

In order to avoid this reaction, after the regeneration of the resin with NaOH, a NaCl may follow which converts the strong base groups to the Cl⁻ form without affecting the weak base groups which can not split salts and therefore remain in the free base form.
Similarly, if the WBA resin is not well rinsed after NaOH regeneration, there will be a leakage of HCOONa at the beginning of the loading cycle.
The WBA resin should have high resisance to physical breakdown due to the swelling of the resin during loading.
For a higher purity of the treated formaldehyde, a SAC➔WBA system can be employed.

Hydrochloric acid purification

One way of HCl synthesis is by reacting sulfuric acid with KCl:

$$H_2SO_4 + 2\ KCl \longrightarrow K_2SO_4 + 2\ HCl$$

The formed HCl gas is absorbed in an absorption column by counter-current flowing water, producing concentrated HCl. This acid contains iron impurities that give an undesired yellow color. When iron is found in the Fe^{3+} form, it forms anionic complexes with HCl such as $FeCl_4^-$, $FeCl_5^{2-}$ and $FeCl_6^{3-}$. These complexes can easily be removed with a SBA resin in the Cl^- form, in a similar way as in the purification of HCl pickle liquors. Regeneration is done with about 3 BV of water. In these cases, Fe^{3+} concentration in the HCl varies from several ppm up to 100-200 ppm. Operating capacities range from several grams Fe/L_R up to 10-15 g Fe/L_R depending on the specific flow rate, the leakage allowed and the Fe^{3+} concentration. Recommended specific flow rate is 2-5 BV/h.

Another source of HCl is as a by-product of the hydrolysis of chlorosilanes to produce silicon polymers. The contaminated HCl may contain silane, siloxane and silanol. The contaminated HCl can be purified from these silicon-containing compounds using a hydrophobic adsorbent resin such as Dowex® XUS-43436 or Dowex® XUS-40323 (Cronin *et al*, 1994). These are almost dry materials having high specific surface area that can adsorb the silicon containing compounds. Regeneration is achieved with a mixture of caustic and a solvent such as alcohols.

Phosphoric acid purification

Phosphoric acid is a triprotic acid with the dissociation constants, $pK_{a1}=2.12$, $pK_{a2}=7.21$ and $pK_{a3}=12.67$. Therefore at acidic pH, phosphoric acid behaves as a monoprotic acid. One of the processes for production of phosphoric acid is the so called wet process. With this process H_2SO_4 is added to the phosphate rock, in the form of fluoroapatite, where the Ca is removed as gypsum by filtration and H_3PO_4 is produced at a concentration 28-30% P_2O_5:

$$Ca_3(PO_4)_2 + 4\ H_3PO_4 \longrightarrow 3\ Ca(H_2PO_4)_2$$
$$3\ Ca(H_2PO_4)_2 + 3\ H_2SO_4 + 6\ H_2O \rightarrow 3\ CaSO_4.6H_2O + 6\ H_3PO_4$$

After filtering off gypsum, phosphoric acid is concentrated through evaporators to reach 35% or 54% P_2O_5 or further to produce "superphosphoric acid" of 70% P_2O_5. Before concentration, it has a concentration of about 5% P_2O_5.

The wet phosphoric acid (WPA) thus produced contains impurities such as sulfuric acid, hydrofluoric acid, hydrofluorosilisic acid, Cd, Cr, Fe, Al, V, Ca, Mg, Na or U. The level of the impurities can be about 10 g/L in the 54% P_2O_5. There exist different techniques to purify WPA from the above impurities such as precipitation, solvent extraction and ion exchange. In the solvent extraction technique, H_3PO_4 is extracted with an organic solvent such as alcohols like n-butyl alcohol, esters like tributylphosphate and similar, and back extracted with water. Ion exchange techniques that are discussed here are applied either on the aqueous phosphoric acid solution or on the non-aqueous solutions.

Removal of cations from H_3PO_4 with a SAC resin in the H^+ form can be achieved on the 5% P_2O_5 acid. Note that 5% P_2O_5 corresponds to 6.9% H_3PO_4 or 0.7 eq/L taking H_3PO_4 as monovalent. A specific flow rate of 4 BV/h is applied and regeneration is done with 5% HCl at a level of 100 g HCl/L_R. Under these conditions, the operating capacity obtained is about 0.8 eq/L_R. Higher concentration H3PO4 can be treated with SAC resins but at lower operating capacities. For example (Dominiani and Annarelli, 1985), 85% H_3PO_4 (61% P_2O_5) containing 11 ppm Ca^{2+} was treated with a conventional SAC resin (Dowex^{TM} 50W-X8) giving an operating capacity of 44 meq/L_R to a 7.9 ppm Ca^{2+} end-point, according to this reference. Resin dehtdration contributes to this low loading capacity.
Vanadium has been removed from 30% P_2O_5 with ion exchange using the following technique (Rossiter et al, 1992). First, vanadium is oxidized to the V^{5+} state. The pentavalent vanadium, VO_2^+, forms anionic complexes in the H_3PO_4 of the type $VO_2.F.H_2PO_4^-$. This solution is allowed to pass through an anion exchange resin, either a WBA or a SBA, where the vanadium complex is fixed. Regeneration is performed by reducing V^{5+} to the V^{4+} state which is then released from the resin since the resin has no affinity for V^{4+}. In order to better remove vanadium from the H_3PO_4, a simulated moving bed system was used in the referenced work.

In another process (Goyden and Hall, 1991) removal of iron, chromium and vanadium from wet process H_3PO_4 is achieved again with a SBA resin. This time, the solution treated is a 15% H_3PO_4 solution in an organic solvent such as tributylphosphate in kerosene. SAC resins instead of the SBA resin were tried in this work but they did not work as well.

Metal impurities such as Cd^{2+} and Cr^{3+} can be removed from WPA using chelating resins with phosphonic functional groups or the Diphonix® resins (Kabay et al, 1998a).
Cadmium removal from 55-65% w/w H_3PO_4 at 90-100 °C has been achieved by complexing Cd^{2+} with halogen ions, specifically with I^- and Br^-, to form anionic complexes, CdI_3^- and CdL_4^{2-}, which are then removed with an anion exchange resin (Tjioe *et al*, 1987). Regeneration was done with dilute (1.4 %) H_3PO_4. the anion resins used resisted chemical degradation after 12 weeks in 70% H_3PO_4 at 90°C.

The removal of uranium from WPA has attracted particular interest because not only it purifies H_3PO_4 for certain applications such as food or fertilizers but the uranium removed can be recovered as strategic material for further use.
Phosphate rock can contain uranium from 100 to 200 ppm, depending on the ore. The removal of uranium from phosphoric acid can be achieved with solvent extraction but also with chelating resin of the aminomethylphosphonic acid type or with inert resins impregnated with di-2-ethyl hexylphosphoric acid (D2EHPA) (Gonzalles-Luque and Streat, 1984; Volkman, 1986; Hassid *et al,* 1986; Kabay *et al*, 1998b). Before passing through the resin, the acid is reduced with metallic iron powder. With this reduction, uranium is converted to the U(IV) state but also Fe(III) is reduced to Fe(II). The reduction of iron is an important feature of the process because Fe(III) poisons the AMP resins (Volkman, 1987). In addition, U^{4+} is fixed stronger than Fe^{2+} so that U^{4+} displaces Fe^{2+} from the resin. In another process (Bristow *et al*, 2013) the iron content of the WPA is previously reduced by precipitation as iron (III) ammonium phosphate, following by reduction of the remaining Fe(III) to Fe(II).

The fixation of uranium by the resin from WPA is a slow process. The operating capacity of the resin depends greatly on the specific flow rate (BV/h) and on the resin particle size. Operating capacities of 6-10 g U/L_R can be obtained.

Elution of the uranium from the resin is achieved with ammonium carbonate. Before elution, the uranium on the resin is first oxidized with raw (unreduced) phosphoric acid, rinsed, washed with ammonia and rinsed.

Rare earths (RE) or rare earth elements (REE) are the 15 lanthanides elements plus scandium and yttrium, usually found in the 3+ oxidation state. WPA represents a secondary source for REE, the primary source being minerals such as bastnasite, monazite and xenotime. With the development of ion exchange technologies, it became possible to separate and isolate RE using SAC resins and selective elutions with EDTA (Woyski, 1965; Winget, 1971).

Rare earths can be recovered from WPA using selective resins with phosphinic functional groups such as Diphonix® resins and Tulsion® CH-96 from Thermax India Ltd (Koumar et al, 2010).

Caustic purification

The purification of NaOH solutions from low levels of NaCl contaminant can be achieved using the ion retardation technique. Thus, by passing a concentrated NaOH solution containing several hundreds of ppm of NaCl through the special retardation Dowex® Retardion® 11A8 resin then rinsing with hot (65°C) water one obtains a caustic solution containing low Cl⁻ levels (The Dow Chemical Company, Caustic Purification with Dowex Retardion 11A8). Figure 5.10 illustrates the leakage profiles for NaOH and NaCl during the loading and the regeneration steps.

In this illustration, the NaOH concentration was 16% and the Cl⁻ level 800 ppm.

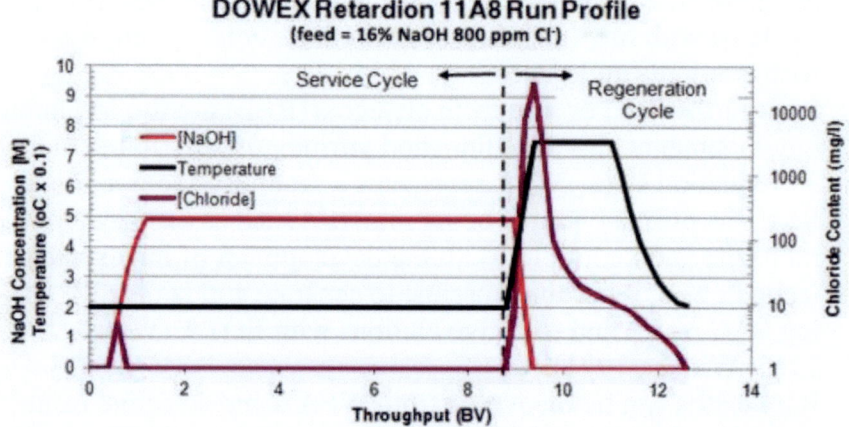

Figure 5.10 Column leakage profiles (From *Caustic Purification with Dowex Retardion 11A8,* The Dow Chemical Company)

Recommended Operation Conditions for Caustic Purification

Operation	Range
Service Flow Rate	1 BV/h
Capacity	~ 8 g Cl⁻/liter resin
Regenerant	65 °C DI Water
Regenerant Volume	2 BV
Regeneration Rate	1 BV/h
Cooling	1 BV, 15°C DI Water

Photographic baths and wastes purification

With the shifting to the digital photography, film devenopments in dark rooms have considerably decreased today.

Nevertheless, the technology of silver removal from photographic effluents
for ecological reasons and recovery for economical reasons is described here.
Silver in photographic rinse waters or wastes comes as thiosulfate complex, mainly $[Ag(S_2O_3)_2]^{3-}$ in ammonium thiosulfate background together with other ingredients such as bleaching agents. Due to the high valency of the silver thiosulfate, anion exchange resins have high affinity for these complexes. Both, weak and strong base resins can be used. Of course, since WBA resins do not split salts, they should be in some salt form.

After resin testing (Chou, 1980) one acrylic WBA resin and one styrenic SBA Type 1 were selected. From these two resins, the WBA gave initially lower operating capacity but regeneration was more efficient: 90% of the absorbed silver was eluted with 6 BV of 30% $(NH_4)_2S_2O_3$ while for the SBA only 60% of silver was eluted with the same quantity of regenerant. It should be noted that the SBA resin had loaded more silver. After 8 cycles, the two resins reached a similar operating capacity, about 20 g Ag/L_R.

Several regenerants have been tested to elute silver from a SBA resin. The more efficient was found to be 30% $(NH_4)_2S_2O_3$ with 12N HCl and 4.7M NH_4Cl next.

The effect of specific flow rate and of the feed solution composition was studied (Mina, 1980). It was found that free thiosulfate in the solution had a big effect on the operating capacity. Specific flow rate, although had an effect at fast flow rates, was less critical. With a feed concentration of 200 mg Ag/L, at 24 BV/h the leakage was 0.45 mg Ag/L.

Synthetic polymeric adsorbents have been used to remove and recycle color developing agents (Bard, 1980). The color developing agents were derivatives of 2,5 diaminotoluene used in various applications such as polymers or ingredients in hair dyes. The product used in this work was 2,5 diaminodiethyltoluene. This product in the free base form in aqueous solution was adsorbed on the synthetic adsorbent Amberlite® XAD4 of Rohm and Haas Company (now The Dow Chemical Company). Elution was performed with H_2SO_4 which converted the product to the protonated form which becomes more water soluble and is removed from the adsorbent. Occasionally, the adsorbent was cleaned with methanol which indicates that the sulfuric acid elution may not remove completely the color developer or that other products were fouling the adsorbent.

6. Chemical synthesis

Fertilizers

Potassium nitrate is produced with the reaction of KCl and some source of nitrates like nitric acid, sodium nitrate or ammonium nitrate. The reaction of nitric acid with potassium chloride:

$$HNO_3 + KCl \leftrightarrows KNO_3 + HCl$$

can be accomplished using a SAC exchange resin in the K^+ form as the source of K^+ ions:

$$R\text{-}SO_3^-K^+ + HNO_3 \leftrightarrows R\text{-}SO_3^-H^+ + KNO_3$$

Continuous systems have been used for a more efficient production, like the Higgins loop (Dennis, 2006) and Advanced Separation Devise (ASD) of Advance Separations Technologies (Abidaud, 1992). The principle of continuous systems is illustrated in figures 6.1 and 6.2.

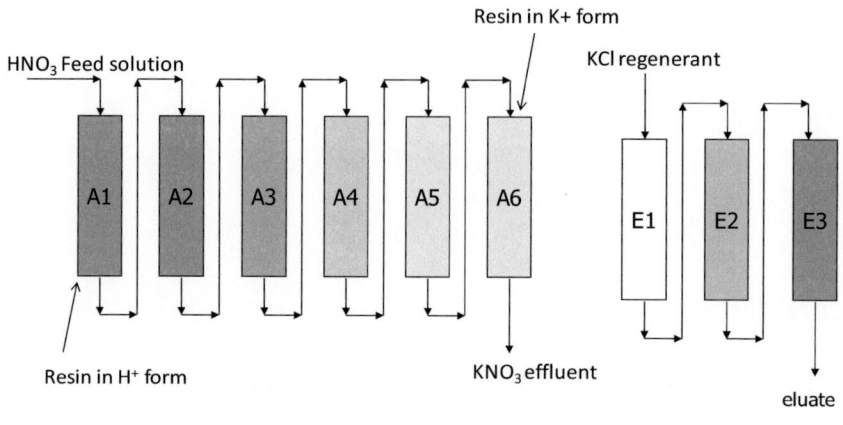

Figure 6.1 Continuous system, principle

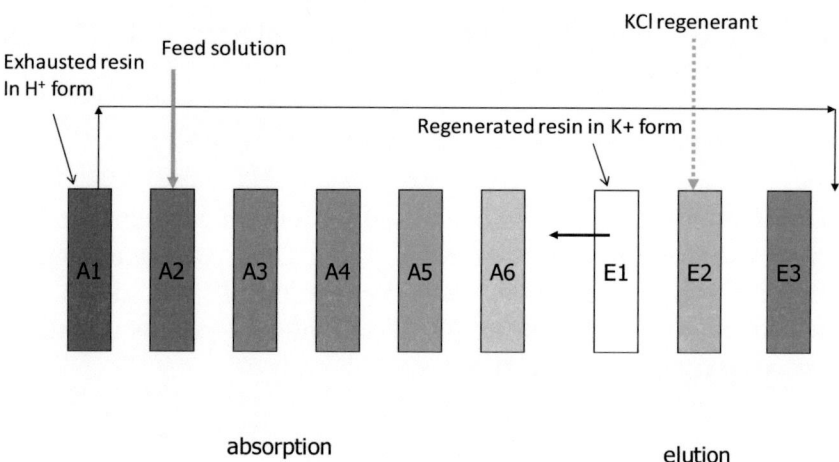

Figure 6.2 Continuous system, resin transfer.

Such a continuous system can be used to produce KNO_3 from HNO_3 and SAC resins in the K^+ form. In practice, there exist different designs of continuous systems, in some of which the resin is periodically transferred to another vessel, or the resin-filled vessels rotate in and out of fixed feed and discharge ports. The ports are arranged in zones where HNO_3 solution, wash water, KCl regenerant and air are supplied continuously. Feed solution consisting of 15% HNO_3 enters the first resin-filled chamber with resin in the K^+ form and exits as a KNO_3+HNO_3 solution. This KNO_3+HNO_3 solution enters the second, adjacent, chamber from which it exits as KNO_3+HNO_3 solution but containing more KNO_3 and less HNO_3 than from the first column. The KNO_3+HNO_3 solution from the second chamber goes to the third one and so on to the last chamber of the HNO_3 zone. The effluent from the last chamber consists essentially of KNO_3 with a low level of HNO_3. This residual HNO_3 can be neutralized with KOH to produce pure KNO_3. The first chamber of the HNO_3 zone, where the resin is by now in the H^+ form, moves in the last position of the water wash zone. Sequentially, this chamber moves progressively to the first position of the water wash zone. At this point, the resin is in the H^+ form and water rinsed. The next zone where the resin-filled chamber moves to is supplied with air where the water filling the chamber is drained off. The following zone is supplied with a 10-12% KCl regenerant solution. The chamber from the first position of the air zone moves then to the last position of the KCl zone. As this resin moves progressively from the last to the first position, it is converted more and more to the K^+ form. Overall, the steps through which each chamber has gone sequentially are HNO_3, water, air, KCl, water, air.

The contact of the resin with HNO_3 entrains the risk of oxidizing the resin with heat release which then favors further oxidation, so that chain reactions may be engaged. To avoid this, appropriate test should be carried out to determine HNO_3 concentration, contact times with resin, temperature and other parameters in order to develop a safe process.

Using the same technique, potassium phosphate can be made in a similar way by reacting $CaHPO_4$ with a K^+ form SAC resin in a continuous system (Berry and Erickson, 1987).

Nitrophosphate fertilizers are products that contain nitrogen and phosphorus at a given $K_2O:P_2O_5$ ratio. They are produced by reacting phosphate rocks with HNO_3. Phosphate rocks react with HNO_3 according to the reaction:

$$Ca_5(PO_4)_3F + 10\ HNO_3 \rightleftarrows 5\ Ca(NO_3)_2 + 3\ H_3PO_4 + HF$$

There is an excess of HNO_3 in the reaction mixture.
In order to manufacture KNO_3 fertilizer, the Ca^{2+} ions should be removed. There exist various processes for that. In the Kemira process, Ca^{2+} is removed with ion exchange (Gowariker et al, 2009). The reaction products are allowed to pass through a SAC resin in the K^+ form where K^+ are exchanged for Ca^{2+}. At the end of the loading cycle the resin is mainly in the Ca^{2+} form. Regeneration is done with KCl:

$$(R\text{-}SO_3^-)_2Ca^{2+} + KCl \rightleftarrows 2\ R\text{-}SO_3^-K^+ + CaCl_2$$

A continuous system as the one described above can be used here as well. The effluent from the resin contains KNO_3, H_3PO_4, excess HNO_3 and some Ca^{2+}. To this solution it is added lime whereby fluorine is precipitated and filtered off and dicalcium phosphate, $CaHPO_4$, (DCP), is produced by increasing the pH and precipitating the DCP. The mother liquor is concentrated and KNO_3 is crystallized. The mother liquor from the crystallizers is recycled back to the ion exchange unit (Suppanen, 2004).

Hydroponics

At this point one can mention ion exchange resins used as fertilizers in hydroponics. Their role consists in supplying nutrients to the water where plants are growing in and at the same time to control the pH and eventually remove free chlorine from water. These resins are usually mixed beds of cation and anion exchange resins loaded with macro- and micronutrients. Macronutrients include N, P, K (primary), Ca, Mg and S (secondary). Micronutrients include elements such as Fe, Mn, Zn, Cu, Cr, Co, Mo and B. The ion exchangers are loaded with appropriate proportions of K^+, Ca^{2+}, Mg^{2+}, NH_4^+ (on the cation exchangers), NO_3^-, SO_4^{2-}, $H_2PO_4^-$, $B(OH)_4^-$ and MoO_4^{2-} (on the anion exchangers). The metals, Fe, Mn, Zn etc, can be loaded either as cations on the cation exchanger or as anionic complexes on the anion exchanger.

As resins, different types have been tried. A mixed bed of a SAC and a WBA has generally been used.

Silica sols

Colloidal silica refers to stable dispersions, or sols, of amorphous silica in water having a particle size varying from 5 to 100 nm and typically, with a uniform particle size. The concentration of silica in the dispersion depends on the particle size.

Thus, silica sols having 50 nm size can be concentrated to 50% by wt while having a size of 10 nm can be concentrated to 30% above which they become unstable. The sols are stable from gelling and settling in a pH range of 8-10.5. They are stabilized with an alkali metal and the particles are negatively charged. Silica sols have many applications including abrasive for polishing silicon wafers, as a coating, in the paper industry used as a drainage aid, as catalyst and others. For these applications, silica sols come in different sizes, size distributions, concentration, pH and surface charge.

Colloidal silica is produced from sodium silicate solutions by partial neutralization. This can be achieved in various ways, by adding an acid or by electrodialysis but the most frequent way is to treat sodium silicate with a hydrogen form cation exchange resin.

There exist two techniques to remove sodium ions from sodium silicate, both using a cation exchange resin. One is a continuous process where a relatively dilute sodium silicate solution is allowed to pass through a fixed bed column of a cation exchanger in the H^+ form (Bird, 1941). The product is an acidic sol which is then stabilized and grown as desired. The other technique is a batch process where the cation exchanger in the H^+ form is agitated batch-wise with the sodium silicate solution in a reactor where the sol particles grow in alkaline pH (Iler and Wolter, 1951).

According to the first process, a dilute (2-6%) sodium silicate solution having a $SiO_2:Na_2O$ ratio of about 3 is allowed to pass through a fixed bed of a cation exchange resin in the H^+ form where Na^+ are fixed on the resin giving an effluent of a very small size (less than 2 nm) of colloidal silisic acid at a pH of 2-4. This colloidal solution is unstable at this pH upon standing. Following the ion exchange resin there are steps where nucleation, polymerization and particle growth take place in the presence of alkali to produce a sol of a pH 8-10.5 and a $SiO_2:Na_2O$ ratio of 20-500.

The sol is subsequently concentrated. The characteristics of the product (size, size distribution, concentration, pH) depend on the process conditions (Yoshida, 2006).

In the second, batch process, a more concentrated sodium silicate solution (10-15%) and a cation exchanger in H^+ form are added to an alkaline reaction medium under agitation where silica particles grow and stabilize.

In the first process, the resin fixed bed, gel type SAC resins are preferred even though macroreticular resins have also been used (Elliott et al, 1976). In the second process WAC resins are preferred but SAC have also been employed.

After the service cycle, the resin is separated from the silica sol by draining, rinsed and regenerated with HCl or H_2SO_4 acid. The regenerant level will determine the fraction of the resin functional groups in the H^+ after regeneration, especially with SAC resins.

The ion exchange process involved for producing silica sols is not a simple ion exchange like for example in a water demineralization process. The characteristics of the produced sols are affected by the rate of exchange of H^+ with Na^+ ions which then depends upon many parameters such as: flow rate (for the fixed bed process), ratio of resin to sodium silicate, rate of addition of resin and sodium silicate to the reactor, agitation rate (for the batch process), temperature, resin characteristics (total capacity, crosslinking level, particle size and particle size distribution, physical porosity), the fraction of the resin groups in H^+ form after regeneration and $SiO_2:Na_2O$ ratio. In order to produce a sol having the desired characteristics of particle size and stability, the exchange of H^+ for Na^+ should therefore be achieved under very tightly controlled conditions.

In addition, the resin should have a good physical stability and stand the periodic acidic and alkaline environments that undergoes between service cycle and regeneration. Otherwise, by breaking, the resin particle size becomes smaller and this changes the kinetics of exchange and eventually the silica sol characteristics. Also the resins should not have any residual organic material remaining from their manufacturing process that may leach out during silica production especially at high temperatures.

Butanol recovery from fermentation broths

Acetone-butanol-ethanol (ABE) fermentation is a bacterial process to produce acetone, butanol and ethanol by fermentation of carbohydrates, developed by Chaim Weitzman at the beginning of the 20th century. This process has been declined with the development of the petrochemical industry but has not been completely abandoned. More recently, improvements have been tried, among others to remove butanol from the fermentation broth in order to reduce the butanol inhibition of the fermentation and a better utilization of the carbohydrates.

Adsorption has been found an interesting approach for butanol removal and recovery from the broth (Qureshi et al, 2005). Synthetic polymeric adsorbents (Amberlite® XAD series of adsorbents) along with polyvinylpyridine, activated carbon and silicalite. Although adsorption gave good capacities for the synthetic adsorbents (Amberlite® XAD 4), desorption and the energy required for that was a key parameter.

7. Waste waters purification

Heavy metals removal from waste waters

There exist various technologies for the removal of heavy metals from waste waters: precipitation as metal hydroxides, carbonates, sulfides or combinations, coagulation-flocculation, adsorption on activated carbon, biomaterials, ferric hydroxides or activated alumina, RO and electrodialysis membranes and ion exchange.
Ion exchange involves conventional cation or anion exchange resins or selective resins to remove metals. The degree of removal of the metals varies depending on the metal, the ionic background, the presence of chelating agents in the solution, the pH, the operating conditions and the type of resin and it can be from a few ppm levels in the feed solutions down to ppb levels or even lower.

Lead, Pb^{2+}, can be found in water from contamination from various sources. Most lead salts are insoluble in water but solubility increases at low pH. Strong or weak acid cation exchangers in Na^+ or H^+ form remove Pb^{2+} from water. Often, the resins act also as a filter due to the presence of (almost) insoluble lead salts such as $PbCO_3$ formed with the alkalinity of the water.

Thiol resins have also a high affinity for Pb and can be used to remove Pb^{2+} from waste water down to very low levels. For example, Ambersep® GT74 treating a water containing 6 ppm Pb at a pH of 2.5, gives a leakage of 0.01 ppm at 15 m/h flow rate (Rohm and Haas, 2006).

$PbCl_2$ is sparingly soluble in water (Clever and Johnston, 1980) but soluble in concentrated HCl forming $PbCl_4^{2-}$. In fact, the solubility of $PbCl_2$ in water is about 8 g/L. In dilute Cl^- background, the solubility decreases due to the common ion effect but above a certain concentration of Cl^- it increases again due to the formation of soluble chloride complexes. In 3N NaCl solution, the solubility of $PbCl_2$ at 20°C is reported to be $17.1*10^{-3}$ mol/dm^3 (3.5 g/L) (Clever and Johnston, 1980). The solubility increases with increasing temperature. Regeneration therefore of the cation exchangers loaded with lead should be investigated using dilute HNO_3, provided that adequate safety care is taken for HNO_3, or high concentration HCl.

Removal of $PbCl_2$ complexes ($PbCl_4^{2-}$) from acidic solutions can be achieved with anion exchange resins. The weak base anion exchangers show higher affinities for Pb(II) anionic complexes than strong base and acrylic anion exchangers showed higher affinities than styrenic (Dabrowski et al, 2004).

The removal of Cr(III) from tannery wastes has been extensively studied (Petruzzelli *et al*, 2002). In one process, the sludges are leached with H_2SO_4 and the solid residues are washed and applied to land. The leachates which contain Fe, Cr(III) and Al(III) are filtered, Fe(II) is oxidized to Fe(III) with H_2O_2 and are treated on IX.

The IX system consists of a formophenolic WBA exchange resin which is used to remove Fe^{3+}. The effluent of the WBA resin is then treated by a SAC resin in the H^+ form. The phenolic

WBA resin (Duolite™ A7 from Rohm and Haas Company, now The Dow Chemical Company) removes Fe^{3+} at a pH of 1.5-2 by forming a chelating complex of $Fe(SO_4)_2^-$ with the O atom of the phenolic group and one N atom. At a pH less than 1.5, Fe is found predominately as $Fe(SO_4)^+$ which is excluded by the protonated resin while at pH>2 Fe is precipitated as iron hydroxides.

The effluents from the formophenolic WBA resin containing Al^{3+} and Cr^{3+} sulfato-complexes were treated by a SAC resin in the H^+ form. The selectivity of the resin for these two metals depends greatly on the ionic speciation at a given pH. Thus, at a pH<1, Cr^{3+} is predominately found as $Cr(SO_4)^+$ while Al^{3+} is found 40% as $Al(SO_4)^+$ and 50% as Al^{3+}. Therefore, at this pH the SAC resin prefers the trivalent Al^{3+} over the monovalent species thus allowing a separation of Al^{3+} from Cr^{3+} which breaks through first.

Contamination of land, waters and air with mercury can happen from various sources like burning fossil fuels, mining, incinerators or industrial production. Mercury-selective resins (thiol, thiourea etc) can be used to remove mercury down to ppb levels. These resins fix ionic mercury only. If mercury is found in the metallic form, $Hg°$, it must first be oxidized to Hg^{2+}. Any oxidant left must be removed before treating the solution of the mercury-selective resins. Typical operating conditions are 10 BV/h specific flow rate. The saturation capacity of the resins depends on the feed solution concentration. For example, thiol resins with a feed Hg^{2+} concentration of 10 ppm have a saturation capacity of about 60-70 g Hg/L_R.

Removal of transition metals from water can be achieved with WBA resins (Höll, 1996) without an ion exchange mechanism, due to the ability of these resins to donate one electron pair of the N atom to a metal at about neutral or alkaline pH to form a coordination complex. It is also possible to use chelating resins having polyamine functional groups to remove metals from waste solutions, as for example, Purolite® S984 to remove cobalt. Manganese can be removed from waste waters using Diphonix® or IDA type resins.

Thallium is a rare metal present as a minor impurity in metal sulfide minerals. Removal and recovery of thallium from industrial wastes has been achieved using thiol resins (Albert and Masson, 1994). The removal of Tl takes place at a pH of 2-5 and at a redox potential so that Tl is in the +1 state. Elution is achieved with 0.5-2 N H_2SO_4. In presence of other metals fixed along with Tl, the elution conditions are chosen, usually the H_2SO_4 concentration, to elute selectively Tl from other elements.
Thallium in its Tl^{3+} form forms a very strong complex with EDTA with a log K_f of 35.3 (Harris, 2011). In fact, it was found that Tl can be removed by IDA or AMP chelating resins (Blondel *et al*, 2012).

Fluorides removal

Fluorides can be removed from waters by adsorption on activated alumina, by aluminum coagulation or by reverse osmosis. An ion exchange technology has been introduced (Popat *et al*, 1994; Oke et al, 2011) using an AMP chelating resin in the Al^{3+} form. This resin gave an operating capacity of about 3-6 g F/L_R depending on the pH of the feed solution. Slightly acidic solutions gave the highest capacity. Specific flow rate was 10 BV/h with a

feed solution containing 127 mg F/L without any other salt background. It is interesting to know that the operating capacity of the resin increases as the salt background of the feed solution inceases. This was explained as due to changes in the ligands around the Al^{3+} atom at different salt backgrounds.
Regeneration of the resin was done with 3 BV of a 5.5% $AlCl_3$ solution.

Arsenic removal

Arsenic is found in water as an anion, either as trivalent, forming arsenous acid $As(OH)_3$ and arsenites, or as pentavalent as arsenic acid, H_3AsO_4 and the arsenate salts. Arsenous acid has pK_{a1}, pK_{a2} and pK_{a3} values of 9.23, 12.13 and 13.40 respectively and therefore in water it is found as non-dissociated at pH below about 9 and for that reason As(III) is more difficult to remove than As(V). Arsenic acid has pK_{a1}, pK_{a2} and pK_{a3} values of 2.2, 7.0 and 11.5 respectively and therefore in water at pH about 6-8 it is found as as $H_2AsO_4^-$ and $HAsO_4^{2-}$.
Arsenic is most frequently removed by first oxidation of As(III) to As(V) followed by coagulation-flocculation process with alum, ferric chloride or ferric sulphate. Other technique include the co-precipitation of As(V) with Fe(III)-oxides where the soluble As(V) is incorporated in the growing Fe(III) hydroxide phase, and adsorption on activated alumina (Al_2O_3), hydeated ferric oxide, activated carbon and other adsorbents.
Adsorbents based on granular ferric hydroxide are also used to remove arsenic, vanadium, selenium, antimony, molybdenum, along with other metals such as chromium, uranium, copper and lead.

Arsenic (III) reacts with thiol groups to form As-S bonds. It is possible therefore to use thiol resins or other supports bearing thiol groups to fix As(III) (Tongesayi, 2014).

Ion exchange resins impregnated with ferric hydroxide have been synthesized for the removal of oxoanions (arsenates, chromates, phosphates, oxalates, phthalates). Although strong acid cation and strong base anions have been used as supports for ferric hydroxide, anion exchange resins showed higher capacity due to the fact that the oxoanions are not excluded by the Donnan phenomenon by the strong base anion exchangers (SenGupta and Cumbal, 2007). Similarly, SBA exchange resins impregnated with hydrated zirconium oxide were synthesized for similar applications (SenGupta and Padungthon, 2013). Arsenic can be removed by SBA resins (Berdal et al, 2000). Since arsenous acid is non-dissociated at pH<9, As(III) must first oxidized to As(V) as arsenic acid, H_3AsO_4 in order to be removed by IX at about neutral pH.

Arseniates and other oxoanions can form complexes with some metals among which Fe^{3+} (see page 35). Thus, chelating resins in the Fe^{3+} form can remove arseniates from water by ligand exchange (Chanda *et al*, 1988 and 1988a). Similarly, phosphonic resins (Diaion® CRP200) in Zr(II) form removes arsenic from water by a similar mechanism (Jyo et al, 2005).

Incinerators wastes treatment

Incinerators burn wastes which they convert to heat, flue gas and ash. The ash consists of the fly ash, which are particles in the flue gas, and bottom ash which is the ash from the incinerator grate. The chemical composition of the flue gas and the ash depends on what the incinerator burns.

During flue gas condensation, the heat of the flue gas is recovered by injected water which is then cooled down to recover heat. The condensed water can then be purified from suspended and dissolved impurities before discharge.

Flue gas may contain impurities such as particulate matter, heavy metals, dioxins, furans, sulfur dioxide, hydrochloric acid and methane. Cleaning the flue gas is done first by removing particular matter (fly ash) by filtration, cyclones or by electrostatic precipitators (ESP).

The removal of gaseous contaminants like HCl, HF, SO_2, is done by injecting powdered chemicals $NaHCO_3$ or $Ca(OH)_2$ which neutralize the acids. These products are subsequently removed by the particular matter removal systems. The removal of these acids can also be achieved by a wet process (scrubbers) using water for HCl and HF followed by a lime scrubber for SO_2. These waters are subsequently treated as waste waters. Activated carbon injection is done to remove organics (dioxins, furans) and volatile metals, Hg and Cd.

Mercury can be removed from wet flue gas desulfurization waste waters down to ppb levels using Hg-selective resins (thiouronium, thiol, thiourea) (Owens *et al*, 2009). The treatment consists of a pretreatment with UF/RO followed by the Hg-selective resin to remove mercury.

The ash, especially the fly ash, may contain heavy metals such as lead, cadmium, mercury, copper and zinc along with small quantities of dioxins and furans. In case of presence of heavy metals, it is possible to leach the ashes with acid to extract the heavy metals and treat the leach liquor on chelating resins.

NH_4NO_3 recovery from wastes of fertilizer plants

Ammonium nitrate fertilizer is produced by direct neutralization of nitric acid with ammonia. After the reaction, the product is concentrated by evaporation. The condensates contain free ammonia and NH_4NO_3 at concentrations about 4-8 g/L ammonia and 6-7 g/L NH_4NO_3.

Free ammonia can first be removed as a gas by for example, pulverizing the condensates and extract ammonia with steam (Leleu, 1983). The remaining can be treated on IER consisting of a SAC and a WBA resin.

The ion exchange system operates according to the Arion process (Arion, 1976 and 1977) as follows:

The IER system consists of a SAC resin followed by a WBA resin. The exchange reactions are:

Cation exchanger:
$$R\text{-}SO_3^- H^+ + NH_4NO_3 \leftrightarrows R\text{-}SO_3^- NH_4^+ + HNO_3$$
Anion exchanger:
$$R\text{-}N + HNO_3 \longrightarrow R\text{-}NH^+NO_3^-$$

The major difference with other processes involving ion exchangers to remove salts from solutions is the regeneration procedure. Here, regeneration of the SAC resin is performed with 45-60% HNO_3 and of the WBA resin with 20-25% NH_3 solution (Arion, 1976).

Cation exchanger:
$$R\text{-}SO_3^- NH_4^+ + HNO_3 \leftrightarrows R\text{-}SO_3^- H^+ + NH_4NO_3$$

Anion exchanger:
$$R\text{-}NH^+NO_3^- + NH_3 \longrightarrow R\text{-}N + NH_4NO_3$$

Because of the use of concentrated HNO_3 acid and the possible resin degradation and eventually explosion, the regeneration conditions are strictly defined. For example, in order to avoid temperature increase during the HNO_3 injection and rinsing, the resin is drained with nitrogen or vacuum, so that there is no HNO_3 dilution with heat release. The regeneration takes place in separate vessel than that of the service cycle. The HNO_3 temperature, flow rate and resin bed depth during regeneration are controlled so that the temperature during regeneration is maintained at levels from 0 to15°C. The resins found to resist oxidation and physical degradation under these operating conditions were defined as the Duolite® C264 and Duolite® A366. They were high DVB content macroreticular SAC resin and polyacrylic macroreticular WBA resin. Styrenic WBA resins may also be considered and used after appropriate testing under the conditions of this process.

Organics removal from wastes

The removal of organic compounds from industrial wastes has been the subject of many investigations since long years due to the discharge of such compounds into the environment and the increasing governmental regulations to lower the levels of the effluent streams. There exist various techniques for reducing the levels of organic compounds in the industrial waste effluents including chemical oxidation, biological degradation and adsorption. Granular activated carbon (GAC) has been used since many years but techniques using synthetic adsorbents have been developed in view of some disadvantages of GAC such as expensive regeneration process and brittleness.

Soon after the development of the macroreticular structure of ion exchangers and the development of macroporous synthetic adsorbents, these products have been used to remove organic compounds from waste waters.

Phenol and phenol derivatives such as chlorophenols, nitrophenols or alkylphenols, were found to be efficiently removed by synthetic adsorbents (Crook *and al*, 1975; Ku and Lee, 2000). Phenols are better removed with these adsorbents at low pH where they are found as neutral molecules. A sharp decrease in operating capacity was observed at high pH where phenol is found as the phenate anion and which is more water soluble. The operating capacity of a conventional adsorbent depends on the phenol concentration in solution. For phenol concentrations in the range 5-6 g/L, capacity is about 60-80 g phenol/L_R. On the other hand, with phenol concentrations in the range of 50-200 ppm the adsorbent capacity is a few to several grams of phenol per liter adsorbent. Leakage in general is less than 1 ppm with specific flow rates of 2-4 BV/h.

Elution of phenols can be achieved with a solvent like methanol or acetone. It can also be achieved, but not as efficiently, with 4% NaOH solution.

High surface area hypercrosslinked adsorbents have also been tried successfully for the removal of phenol derivatives (Oh *et al*, 2003). Higher loading capacities compared to conventional, first generation, adsorbents were experienced due to their higher specific surface areas.

Phenolic compounds were found to be removed from aqueous solutions using anion exchange resins (Chasanov et al, 1956; Ku and Wang, 2005). Phenol is a weak acid and as such, it can be removed from aqueous solutions at alkaline pH using a SBA resin in the OH form. Regeneration can be achieved with 4% NaOH.

Aromatic compounds such as benzene, toluene and xylene (BTX) are easily removed by synthetic adsorbents of styrene-DVB type having small size pores. From a feed solution containing 200-300 ppm of BTX, the treated effluent contains less than 1 ppm at a flow rate of up to 10 BV/h. The loading capacity, depending on the adsorbent, is about 30 g/L_R. Steam regeneration can be applied. If there is high level of xylene, then high pressure steam would be necessary.

Ethylene dichloride removal from wastes can be done with similar adsorbents as in the case of BTX. Using a lead-lag installation, from a feed solution containing 8000 ppm EDC an effluent containing < 2 ppm can be obtained. Steam regeneration can be applied at 120°C.

8. Air purification

The purification of air with ion exchangers is a technique used since many years and was recently reviewed (Soldatov and Kosandrovich, 2011).
HCl removal from heavily contaminated air can be achieved with wet scrubbing. However, to decrease the concentration to below 100 mg/m3 one technique involves SBA resins in the OH^- form.
Ammonia removal from air was done with SAC resins in the H^+ form. The efficiency of the removal depends on the humidity in the air. Humid resin removes NH_3 from air much faster than dry resin. CO_2 can be removed by SBA resin in the OH^- form. Regeneration is performed with NaOH solution.

There exist several techniques to remove volatile organic compounds (VOC) from gas streams. These include incineration, scrubbing, adsorption or condensation with possible phase separation. The adsorption technique is discussed here with reference to synthetic polymeric adsorbents and to carbonaceous adsorbents, specifically those made by pyrolysis of macroreticular SAC exchange resins under the trademark Ambersorb® by Rohm and Haas Company (now part of The Dow Chemical Company) (Maroldo et al, 1990).

During treatment of gas streams containing organic vapors, the gas stream passes through a bed of synthetic or carbonaceous adsorbent, either fixed bed or fluidized bed, and the organic contaminants are removed from the gas stream and fixed on the adsorbent. When the adsorbent becomes exhausted, it is regenerated by passing a heated inert gas, for example nitrogen, through the adsorbent. The inert gas exits the column along with the organic contaminants and are directed to a condenser where the organics are recovered. It is conceivable that the exhausted adsorbent is regenerated with steam. Steam is a more concentrated source of heat than hot air so it can more quickly and efficiently regenerate the adsorbent. However, the wet adsorbent does not adsorb VOC as efficiently as the dry and therefore it may be necessary to cool and dry the adsorbent before the following loading cycle.

The partial pressure of the organic vapor has an important effect on the adsorption capacity of the adsorbent. Therefore, the concentration of the organic vapor in the gas (ppm) and the pressure of the gas are important parameters. Other important parameters are temperature, velocity of the gas stream and relative humidity (RH).

The concentration of VOC in a gas stream is usually expressed as mg VOC per kg of gas (ppm). Taking a composition for the carrier gas (air) and the corresponding molecular weights, the concentration of the VOC can be converted to mole fraction. For example, if the concentration of a VOC in a gas stream is α mg/kg of air, then the number of moles in a kg of air is $\alpha/(MW*1000)$ where MW is the molecular weight of the VOC. From similar calculations for all components of the gas stream, one can calculate the mole fractions of the components. Given the pressure, the vapor pressures of the individual components can be calculated from the expression (Vapor pressure) = (pressure)*(mole fraction). The saturation capacity of an adsorbent for various compounds can be determined by constructing an isotherm, at given conditions of temperature and RH, where the saturation capacity of the adsorbent is plotted against the vapor pressure of the compound. From the calculated vapor pressure one obtains the saturation capacity of an adsorbent for each of the components of the gas stream. The overall saturation capacity is taken as the sum of $Q_i*c_i/\Sigma c_i$ where Q_i is the individual saturation capacity and c_i is the individual component concentration. Assuming that only a fraction of the saturation capacity is the operating capacity, one can calculate the quantity of adsorbent needed to clean the gas stream.

Temperature has a negative effect on the equilibrium capacity of the adsorbent. Increasing temperature reduces the equilibrium capacity. However, temperature has a positive effect on the kinetics of adsorption and therefore the overall effect may be positive or negative.

Relative humidity has a negative effect on adsorption. Increasing RH decreases the equilibrium capacity of the adsorbent because water will take up sites where the organic vapor will no longer be adsorbed. This effect is even more negative if the adsorbent is already wet before starting the service cycle with the humid gas stream. For that reason, as pointed out earlier, after steam regeneration, the adsorbent is dried before starting the service cycle.

9. Non-aqueous systems

The use of ion exchange in non-aqueous systems was already suggested in 1949 with a special resin, Amberlite™ IRC50, at times where only gel-type ion exchange resins were available and which in most cases failed to work in organic solvents (Rohm and Haas Company, 1949).

IER beads are imbibed with a solvent, most frequently water. This water is of a primary importance since all the ion exchange reactions take place in solution.

If a dry or partially dry resin comes in contact with water, the water from outside the resin beads will enter into the resin and the resin will swell (Helfferich, 1962). This swelling results from the tendency of the hydrophilic fixed and mobile ionic groups to become hydrated making the polymer chains to expand. Also, since the ionic concentration inside the resin is higher than the ionic concentration in the external solution, water enters into the resin to balance the osmotic pressure difference between the interior of the resin and the external water. A third reason for the swelling of the resin is the electrostatic repulsion of fixed neighbouring ionic groups. The swollen resin will approach equilibrium with the outside water as the above stretching forces (hydration of ions, osmotic pressure difference and the increasing distance of neighbouring ionic groups) decrease. In general, the swelling of a resin in a solvent depends upon the crosslinking density of the polymer matrix, the nature of the polymer (aromatic, acrylic, formophenolic), the nature and the concentration of the ionic functional groups (the total exchange capacity) and the nature of the counter-ions.

In case of an organic solvent, the resin behaves in that solvent in a similar way but naturally, the extent depends on the properties of the solvent (Trémillon, 1965). Thus, the swelling of the resin depends on the affinity of the solvent molecules for the fixed ionic groups as well as for the mobile ions. This affinity depends essentially on the polarity of the solvent. For example, sulfonic resins swell more in water than in most of the organic solvents due to the higher affinity of water molecules for the sulfonic groups. Similarly, the hydration number of the alkali metals is higher than the solvation number in alcohols for example, resulting in a different size of the alkali metals counter-ions in water and in alcohols. The size of Li^+ in water is larger than the rest of the alkali metals and therefore a sulfonic resin in water swells more in the Li^+ form than in any other alkali metal form. In alcohol on the other hand it is the opposite, the solvated Li^+ form resin swells less than with the other alkalis.

In addition to this effect of affinities of the solvent molecules for the fixed or mobile ionic groups of the resin, the dielectric constant of the solvent plays an important role on the dissociation and therefore on the swelling of the resin. Thus, the lower the dielectric constant, the stronger is the attraction of the ions of different charge with the result that the dissociation of the electrolyte is less and the swelling will be less. For the same reason, the repulsion of the fixed ionic groups will be bigger, the distance between these fixed ionic groups will be larger and the swelling more.

Ion exchange equilibria are affected by the solvent (Helfferich, 1962). Degree of dissociation, ionic solvation, complex formation are parameters that affect equilibrium. For example, cyano complexes of gold can be eluted from SBA resins with HCl in aqueous acetone but not with aqueous HCl solution. Also, alcoholic solutions favour the uptake of metal chloro-complexes by SBA resins compared to aqueous solutions.
In mixed miscible solvents in contact with a resin, both solvents enter the resin beads but in different proportions than in the external solution. The higher the affinity of one constituent of the mixed solvents for the resin, the higher will be the proportion of this constituent in the resin. The distribution of the constituents of the solvent between resin and external solution depends on their solvating power for the fixed as well as for the mobile ionic groups.

An important case is when the organic solvent is immiscible with water. In general, the IER after the service cycle are regenerated with aqueous solutions of an acid or base. Therefore, the regeneration procedure is adapted accordingly. If a resin swollen with water comes in contact with a solvent which is immiscible with water, for example benzene, then the water will stay in the resin beads and no penetration of the organic solvent (benzene) into the resin will take place. In order to make the external solution enter the resin beads, a third solvent is used, for example alcohol, which is miscible with both, water and benzene. First water is displaced from the resin with the alcohol, followed by the displacement of alcohol by benzene. This step is called sometimes "sweetening on". In order to rewet the resin, then alcohol is introduced to displace the benzene, followed by water to displace alcohol. This is the sweetening off step, using the same term as in aqueous applications where after the service cycle follows a water displacement step to displace the treated solution before injecting the regenerant. This procedure is used in a process where the IER operates in the service cycle in an immissible solvent while the regeneration takes place in water solutions. The sweetening off and sweetening on steps preceed and follow the regenerant injection.

In the following, it is discussed the use of IER in applications where the solvent is not water or water is a minor constituent.

Phenol deacidification

The production of phenol is based on the cumene process according to which benzene and propylene react to give cumene which subsequently is oxidized to cumene hydroperoxide which in acidic conditions gives phenol and acetone.

PhC(OOH)(CH₃)₂ →[H₂SO₄] PhOH + (CH₃)₂C=O

The two final products are recovered by distillation, however, beforehand it is desirable to remove the H_2SO_4 to avoid color formation and other by-products. This can be achieved with a WBA resin in the free base form or a combination of a SBA and a WBA resin to remove any other weak acid impurities present in the cleavage mixture.

The system used frequently is the three column merry-go-round where two columns are under loading in series while the third is on regeneration. A specific flow rate of 3-4 BV/h (1 BV is the resin volume of one column) can be employed. Temperature can vary from ambient to 50°C. With a feed concentration of 100-150 ppm H_2SO_4, an operating capacity of about 35 g H_2SO_4/L_R can be obtained.

Because phenol is not miscible with water, a sweetening off and sweetening on step between loading and regeneration with acetone is necessary. Thus, the regeneration steps are exhaustion, sweetening off with 2-3 BV of acetone, water displacement, caustic regeneration at 4 BV/h for 30-45 minutes, water rinse, sweetening on with 2-3 BV of acetone followed by the introduction of the feed stream. Styrenic, phenolic, acrylic or pyridine type WBA resins have been suggested (Cipullo, 1992). Because the feed solution may still contain some peroxides, the ion exchange resins may be oxidized and show reduced life time, especially the phenolic type.

After being distilled, phenol still contains impurities that are difficult to be separated from phenol due to the similar molecular weight. These impurities include methylstyrene, cumene, acetophenone, mesityl oxide. A method to reduce these impurities is to convert them to easily separable compounds using SAC resins as catalysts. These impurities are condensed with phenol to make high boiling products.

Similar deacidification process as the one described above can be used to deacidify the purified phenol from acidic impurities such as oligosulfonates coming from the SAC resin catalyst. In the production of bisphenol A from phenol and acetone with a SAC resin as catalyst, the product bisphenol A with the unreacted phenol contains some acidic impurities. After separating from the solid acid catalyst, the mixture can be treated with a WBA resin of pyridine type to purify from the acidic impurities (Iimuro et al, 1992).

Methanol purification

Already in 1949 a weak acid cation exchanger, Amberlite™ IRC50, was suggested to remove amines from alcohols (Rohm and Haas Company, 1949). Methanol purification from metal impurities Mn^{2+} and Fe^{3+}, was achieved with a gel-type SAC resin in the H^+ form (Tidwell, 1957). If crude methanol is neutralized and distilled in the presence of these metal ions, a fine precipitate is obtained difficult to filter off. To avoid this, crude methanol is treated with a SAC resin in H^+ form at a specific flow rate of 16 BV/h. The effluent is then neutralized with NaOH and distilled to recover purified methanol. Regeneration of the resin can be achieved with a 10% H_2SO_4 solution.

Methanol is purified from amine impurities using a macroreticular SAC resin in the H^+ form at a specific flow rate of about 10 BV/h. Since the levels of the amine impurities are low (< 1ppm), and depending on the quantity of methanol needed to purify, the resin can be used only once and replaced when exhausted.

Mercury removal from hydrocarbons

There exist various technologies for the removal of mercury from hydrocarbons. In general, they consist of an adsorbent like zeolites, activated carbon, metal oxides such as alumina, containing an active compound for mercury removal, like sulfur, silver, KI, or metal sulfides.
Mercury-selective resins having functional groups such as thiol, thiourea or thiouronium have been used to remove mercury from various solutions or waste waters. In these aqueous systems mercury is removed only in the ionic form. The use of mercury-selective ion exchange resins containing thiol or thiourea groups to remove mercury from non-polar solutions has been reported (Duisters and van Geem, 1990). It was found that these resins containing thiol groups, -SH, or thiourea that gives thiol groups via tautomerization, can remove mercury from non-polar liquors. The condensate of a gas stream with hydrocarbons higher than C_4 is allowed to pass through the ion exchange material containing thiol or thiourea groups at a flow rate of 30-60 BV/h and ambient temperature. Thiol containing resins such as Ambersep® GT74 of The Dow Chemical Company can be regenerated with concentrated HCl.

Biodiesel purification

Biodiesel is produced by transesterification of fats and oils, usually with methanol: the fat reacts with methanol to form methyl esters and glycerol, which is separated from the methyl esters by phase separation:

$$\begin{array}{c} CH_2OCOR_1 \\ | \\ CHOCOR_2 \\ | \\ CH_2OCOR_3 \end{array} + CH_3OH \xrightarrow{\text{catalyst}} \begin{array}{c} R_1COOCH_3 \\ R_2COOCH_3 \\ R_3COOCH_3 \end{array} + \begin{array}{c} CH_2OH \\ | \\ CHOH \\ | \\ CH_2OH \end{array}$$

Oils, fats   Fatty acids esters   glycerol

The glycerine phase can be purified with ion exchange as described elsewhere (Zaganiaris, 2011, p. 187).
Most frequently the reaction uses homogeneous alkaline catalysts, NaOH or KOH. After the separation of the glycerine phase, crude biodiesel contains small quantities of glycerol and soaps, formed by the reaction of fats with the NaOH or KOH catalysts.

$$\begin{array}{c} CH_2OCOR_1 \\ | \\ CHOCOR_2 \\ | \\ CH_2OCOR_3 \end{array} + NaOH \xrightarrow{\text{catalyst}} \begin{array}{c} R_1COONa \\ R_2COONa \\ R_3COONa \end{array} + \begin{array}{c} CH_2OH \\ | \\ CHOH \\ | \\ CH_2OH \end{array}$$

Oils, fats   soaps   glycerol

Other impurities that crude biodiesel can contain are free fatty acids, water, methanol, catalyst and glycerides.

One way to remove the residual glycerol and soaps from biodiesel is by water wash. Biodiesel is mixed with water, agitated and then allowed to separate the two phases where glycerol and soaps are removed with the water. The water wash presenting some drawbacks like biodiesel lost, emulsion formation or phase separation, there are alternative ways to purify biodiesel including with ion exchange.

The ion exchange resins used are SAC resins in the H^+ form, in the dry state. The dry resin removes the cation from the soap impurities (Na^+ or K^+), generating free fatty acids in the biodiesel, residual catalyst, NaOH or KOH, and glycerine, by adsorption and/or filtration. There is a synergistic effect with some soap being adsorbed together with glycerine (Wall, 2009). After the ion exchange resin, biodiesel is dried to remove the methanol.

REFERENCES

Abidaud A (1992). Continuous production of potassium nitrate via ion exchange. US Patent 5,110,578.

Abisheva ZS, Zagorodnyaya AN (2011). Rhenium of Kazakhstan (Review of technologies for rhenium recovery from mineral raw materials in Kazakhstan). *In German et al 7th International Symposium on Technitium and Rhenium-Science and Utilization, July 4-8, 2011.* Proceedings, Granitsa, Moscow, 2011, 208-216.

Ackermann F, Berrebi G, Dufresne P, van Lierde A, Foguenne M (1993). Recovery of molybdenum and vanadium from used catalysts. *European Patent EP0555128.*

Ahmad R, Ahmad N, Malik MA, Wahid A (1994). Metal complexes of some phenolic 1,3 diketones II: Mononuclear complexes of some bivalent metal ions with 1(2-hydroxyphenyl)-3(4-halophenyl)-1-3 propanediones. *J Chem Soc Pak* **16**: 216-220.

Albert L, Masson H (1994). Thallium extraction process. *US Patent 5,296,204.*

Alexandratos SD, Chiariza R, Gatrone RC, Horwitz EP (1994). Phosphonic acid based ion exchange resins. *European Patent EP0618843 A4.*

Alexandratos SD, Brown GM, Bonnesen PV, Moyer BA (2000). Bifunctional anion-exchange resins with improves selectivity and exchange kinetics. *US Patent 6,059,975.*

Alfaro E, Frenay J (2004). Recovery of gold from alkaline thiosulfate solutions with ion exchange resins. *In* M. Cox, *Ion Exchange Technology for Today and Tomorrow, IEX 2014,* SCI, 141-148.

Alguacil FJ, Adeva P, Alonso M (2005). Processing of residual gold(III) solutions via ion Exchange. *Gold Bulletin* **38(1):** 9-13.

Araki CD, Ciminelli VST, Freitas LR (2004). Ammonium thiosulfate leaching of an oxide gold-copper ore. *XX Encontro Nacional de Tratamento de Minérios e metallurgia Extrativa*, Florianópolis, v.1, 343-350.

Arden TV (1968). *Water Purification by Ion Exchange.* Butterworth, London.

Arion NM (1976). Process for the regeneration of ion exchange resins and applications thereof. *US Patent 3,956,115.*

Arion MN (1977). Process for treating and recovering waste water from the fertilizer manufacture. *US Patent 4,002,455.*

Arrighi R, Pastacaldi A (1998). Process and installation for the purification of an aqueous solution of an alkali metal chloride. *European Patent 0659686.*

Aylmore MG (2016). Alternative Lixiviants to Cyanide for Leaching Gold Ores. *In* Adams MD *Gold Ore Processing Project Development and Operations* 2nd Edition, Elsevier 2016 p. 447-484.

Bäcklund LY, Rennerfelt LI (1981). Purification process for spent pickling baths. *EP 0035515 A1*

Bailey C, Harris GB, Kuyvenhoven R, du Plessis J (2005). Removal of nickel from cobalt sulphate electrolyte using ISEP® continuous ion exchange. *Calgon Carbon Corporation.*

Bard CC (1980). Recovery and reuse of color developing agents. *Soc Motion Picture and Television Engineers J* **89**: 225-228.

Bauman WC (1954). Separation of substances having different degree of ionization. *US Patent 2,684,331*

Berdal A, Verrié D, Zaganiaris E (2000). Removal of arsenic from potable water by ion exchange. *In* Graig JA Ion Exchange at the Millenium. Imperial College Press, 101-108.

Berdal A, Rezkallah A, Zaganiaris E (2004). The use of ion exchange resins in brine purification in the chlor-alkali membrane cell process. *Presented at the annual conference of the Chinese Chlor-alkali Association, Nankin, China, May 20, 2004.*

Berry WW, Erickson WR (1987). Production of potassium phosphates by ion exchange. *US Patent 4,704,263.*

Bird GP (1941). Colloidal solutions of inorganic oxides. US Patent 2,244,325.

Blokhin AA, Maltseva EE, Pleshkov MA, Murashkin JV, Mikhaylenko MA (2011). Sorption recovery of rhenium from acidic sulfate and mixed nitrate-sulfate solutions containing molybdenum. *In* German et al *7th International Symposium on Technitium and Rhenium-Science and Utilization, July 4-8, 2011.* Proceedings, Granitsa, Moscow, 2011, 254-261.

Blondel JM, Grosjean F, Humblot C, Nicolas F (2012). Method for recovery of thallium from an aqueous solution. *Patent WO 2012143394 A1.*

Bolden WB, Groves Jr FR (1990). Amine recovery by ligand exchange: pore diffusion model. *Ind Eng Chem Res* **29:** 116-121.

Bolto BA, Cope AFG, Stephens GK, Weiss DE (1976). Advances in thermally regenerated ion exchange. *The theory and practice of ion exchange,* An International Conference, Cambridge, UK, July 25-30, 1976. Society of Chemical Industry. 29.1-29.11

Boryta DA, Donaldson AJ (2012). Production of high purity lithium compounds directly from lithium containing brines. *EP 2487136A1*

Brintzinger H, Hester RE (1966). Vibrational analysis of some oxoanion-metal complexes. *Inorg Chem* **5:** 980-985.

Bristow NW, Chalmers MS, Davidson JA, Jones BL, Kucera PR, Lynn N, Macintosh PD, Page JM, Pool TC, Richardson MW, Soldenhoff KH, Taylor KJ, Wayrauch C (2013). Extraction of uranium from wet-process phosphoric acid. *US Patent Application 20130022519 A1.*

Brown CJ, Dejak MJ (1987). Process for removal of copper from solutions of chelating agent and copper. *US Patent 4,666,683*

Brown CJ (1997). Mixed acid recovery with the APU™ acid sorption system. *Eco-Tec Technical Paper N° 147.*

Brown CJ, Sheedy M, Paleologou M, Tompson R (1998). Ion exchange technologies for the minimum effluent Kraft Mills. *Presented at the CPPA Technical Section Symposium on System Closure II. Montreal, January 26-30 1998.*

Brown CJ (2002). Recovery of stainless steel pickle liquors: Purification vs Regeneration. *Eco-Tec Technical Paper N° 158*

Brown CJ (2007). Short-bed ion exchange. *In*: Sengupta AK *Ion Exchange and Solvent Extraction, A Series of Advances Vol. 18,* CRC Press, 375-403.

Broz S (1973). Process for preparing monoethylene glycol and ethylene oxide. *US Patent 3,904,656*

Byler RE, Dunn RC (1953). Process for the recovery of precious metal values. US Patent 2,648,601.

Bywater N (2011). The global viscose fiber industry in the 21st century-the first ten years. *Lenzinger Berichte* **89:** 22-29.

Carr J, Chamberlain T, Zontov N (2012). Method and system for extraction of uranium using an ion-exchange resin. *Patent application WO 2012109705 A1.*

Chanda M, O'Driscoll KF, Rempel GL (1988). Ligand exchange sorption of arsenate and arsenite anions by chelating resins in ferric ion form. I. Weak base chelating resin Dow XFS-4195. *Reactive Polymers* **7:** 251-261.

Chanda M, O'Driscoll KF, Rempel GL (1988a). Ligand exchange sorption of arsenate and arsenite anions by chelating resins in ferric ion form. II. Iminodiacetic chelating resin Chelex 100. *Reactive Polymers* **8:** 85-95.

Chasanov MG, Kunin R, McGarvey F (1956). Sorption of phenols by anion exchange resins. *Ind Eng Chem* **48:** 305-309.

Cheng GW, Crisosto CH (1997). Iron-Polyphenol complex formation and skin discoloration in peaches and nectarines. *J Amer Soc Hort Sci* **122 (1):** 95-99.

Chiariza R, Horwitz EP, Alexandratos SD, Gula MJ (1997). Diphonix® resins: A Review of its Properties and Applications. *Separation Sci Technol* **32:** 1-35.

Chou HW (1980). Silver recovery from wash water with ion exchange. *J Appl Photographic Eng* **6:** 14-18.

Cipullo MJ (1992). Removal of acids from phenol using anionic exchange resins. *US Patent 5,124,490*

Clever HL, Johnston FJ (1980). The solubility of some sparingly soluble lead salts: An evaluation of the solubility in water and aqueous electrolyte solution. *J Phys Chem Ref Data* **9:** 751-784.

Courduvelis CI, Gallager GC (1981). Selective removal of copper or nickel from complexing agents in aqueous solution. *US Patent 4,303,704*.

Cox H, Schellinger AK (1958). An Ion Exchange Approach to molybdic oxide. Eng Min J **159:** 101
Cronin KA, Evanko WA, Malekadeli A (1994). Purification of hydrochloric acid. *US Patent 5,330,735*.

Crook EH, McDonnell RP, McNaulty JT (1975). Removal and recovery of phenols from industrial waste effluents with

Amberlite XAD polymeric adsorbents. *Ind Eng Chem Prod Res Dev* **14:** 113-118.

Dabrowski A, Hubicki Z, Podkoscielny P, Robens E (2004). Selective removal of the heavy metal ions from waters and industrial waste waters by ion-exchange method. *Chemosphere* **56:** 91-106.

D'Alelio GF (1954). Ion-exchange resins from a vinyl heterocyclic amino compound and a vinyl cyclohexene. US Patent 2,683,124.

Dai X, Simons A, Breuer P (2012). A review of copper cyanide recovery technologies for the cyanidation of copper containing gold ores. *Minerals Engineering* **25:** 1-13.

Davankov V, Tsyurupa M, Ilyin M, Pavlova L (2002). Hypercross-linked polystyrene and its potentials for liquid chromatography : a mini-review.*J Chromatography A* **965:** 65-73.

de Dardel F and Arden TV (1989). Ion Exchangers. *In Encyclopaedia of Technical Chemistry.* VCH, Weinheim, Germany Vol. A14

Dennis RS (2006). Continuous ion exchange for fertilizer and phosphate applications. AIChE Central Florida Conference, Sand Key, FL.

Devos C, Demay D, Dulphy H (2001). Process for purification of aqueous solutions of hydrogen peroxide. *European Patent EP 1095905.*

Diniz CV, Doyle FM, Ciminelli VST (2002). Effect of pH on the adsorption of selected heavy metal ions from concentrated chloride solutions by the chelating resin Dowex M-4195. *Sepaeation Sci and Technol* **37:** 3169-3185.

Dominiani FJ, Annarelli DC (1985). Treatment of concentrated phosphoric acid. *US Patent 4,551,320.*

The Dow Chemical Company (1960). Ion Retardation. *Chem Eng News* **38:** 64-65.

The Dow Chemical Company. Dowex® M4195 Product information. *Form number 177-01817-0306.*

The Dow Chemical Company. Ion Retardation: Dowex® 11A8 Retardation Resin.

The Dow Chemical Company. Caustic Purification – Reducing Chloride Levels in Caustic with DOWEX Retardion 11A8 Ion Retardion Resin

Dreisinger DB, Leong BJY (1994). Method for selectively removing antimony and bismuth from sulfuric acid solutions. *US Patent 5,366,715.*

Duisters HAM, van Geem PC (1990). Process for removing mercury from a non-polar organic medium. US Patent 4,950,408.

Dungan LJ, Han L (2001). Sludge-free treatment of copper CMP wastes. *US Patent 6,306,282.*

Duyvesteyn WPC, Neudorf DA, Weenink EM (2002). Resin-in-pulp method for recovery of nickel and cobalt. *US Patent 6,350,420.*

E&MJ News, Engineering and Mining J (2010). Resin-in_solution approach solves gold-copper selectivity problem. Published March 10, 2010.

Edwards RI, Haines AK, Te Riele WAM (1976). The separation of gold from acidic leach liquors with Amberlite XAD-7. *The theory and practice of ion exchange,* An International Conference, Cambridge, UK, July 25-30, 1976. Society of Chemical Industry. 40.1-40.12.

Elliott Jr CH, Hoffman GH, Nozemack RJ (1976). Fluid cracking catalyst based on silica sol matrix. *US Patent 3,972,835.*

Fernandez-Olmo I, Ortiz A, Urtiaga A, Ortiz I (2008). Selective iron removal from passivating baths by ion exchange. *J Chem Techn and Biotechnology 83, 1616-1622.*

Fleming CA, Hancock RD (1979). The mechanism in the poisoning of anion-exchange resins by cobalt cyanide. *J S Afr I Min Metall* **79:** 334-341.

Fleming CA, Cromberge G (1984). The extraction of gold from cyanide solutions by strong- and weak-base anion-exchange resins. *J S Afr Inst Min Metall,* **84**: 125-137.

Fleming CA (1985). The regeneration of thiocynate resins. *S.Afr.Pat. 84/0244.*

Fleming CA (2003). The economic and environmental case for recovering cyanide from gold plant tailings. *SGS Minerals Services Technical Bulletin 2003-02.*

Fleming CA (2005). Cyanide recovery. *In*: Adams ND *Advances in Gold Ore Processing.* Elsevier BV, Editor, Part II, Ch. 29, 703-728.

Fradkin AM, Tooper EB (1955). Treatment of spent sulfuric acid pickling liquors. *Ind Eng Chem* **47:** 87-90.

Gaikwad RW, Gupta DV (2008). Review of removal of heavy metals from acid mine drainage. *Appl Ecology and Environ Res* **6:** 81-98.

Gedgagov EI, Nekhoroshev NE (2011). Use of ion exchange resins for producing high purity ammonium perrhenate in processing of crude ore and manmade rhenium-containing raw materials. *In* German et al *7th International Symposium on Technitium and Rhenium-Science and Utilization, July 4-8, 2011.* Proceedings, Granitsa, Moscow, 2011, 218-221.

Geffen N, Semiat R, Eisen MS, Balazs Y, Katz I, Dosoretz CG (2006). Boron removal from water by complexation to polyol compounds. *J Membrane Sci* **286:** 45-51.

Goldblatt E (1959). Recovery of cyanide from waste cyanide solutions by ion exchange. *Ind Eng Chem* **51:** 241-246.

Gonzallez-Luque S, Streat M (1984). The recovery of by-product uranium from wet process phosphoric acid solutions using selective ion exchange resins. *In* Naden D, Streat M *Ion Exchange Technology* Ellis Horwood Limited, 679-689.

Goodall WR, Leatham JD, Scales PJ (2005). A new method for determination of preg-robbind in gold ores. *Minerals Engineering* **18:** 1135-1141.

Goyden DD, Hall RE (1991). Process for removing iron, chromium and vanadium from phosphoric acid. *US Patent 5,006,319.*

Greager IP, Wyethe JP, Kotze MH, Dew D, Miller D (2001). A resin-in-pulp process for the recovery of copper from bioleach

CCD underflows. *Copper Cobalt Nickel and Zinc Recovery Conference, The South African Institute of Mining and Metallurgy in collaboration with The Institution of Mining and Metallurgy (Zimbabwe Branch), Victoria Falls, Zimbabwe.*

Green BR, Potgeiter AH (1984). Unconventional weak-base anion exchange resins, useful for the extraction of metals, especially gold. *In* Naden D, Streat M *Ion Exchange Technology*. Ellis Horwood Limited, 626-636.

Green BR, Tyc I, Schwellnus AH (1992). Gold selective ion exchange resins. *US Patent 5,134,169*.

Green BR, Smit DMC, Maumela H, Coetzer G (2004). Leaching and recovery of platinum group metals from UG-2 concentrates. *J South African Institute of Mining and Metallurgy* **104**, *323-332*

Grinstead RR (1979). Copper-selective Ion-exchange resin with improved iron rejection. *J Metals* **31 (3):** *13-18*.

Gu B, Brown GM, Alexandratos SD, Ober R, Patel V (1999). Selective anion exchange resins for the removal of perchlorate ClO4- from ground water. *ORNL-TM 13753 report.*

Guo WJ, Shen YH (2015). Recovery of molybdenum and vanadium from acidic sulfate leach solution of blue sludge by ion exchange. *Environ Prog Sustainable Energy doi:10.1002/ep 12220.*

Gupta CK, Mukherjee TK (1990). *Hydrometallurgy in Extraction Processes Vol. II, CRC Press, 99-*

Haines AK, Tunley TH, Te Riele WAM, Cloete FLD, Sampson TD (1973). The recovery of zinc from pickle liquors by ion exchange. *J South African Inst of mining and metallurgy, Nov. 1973, 149-157.*

Harris BG, Barry J-P, Monette S (1991). Recovery of gold from acidic solutions. *US Patent 5,028,260.*

Harris B, White C (2012). Process for the recovery of gold from an ore in chloride medium with a nitrogen species. *European Patent EP 2,536,860 A1.*

Harris DS (2011). *Quantitative Chemical Analysis, 8th Edition, Chapter 11,* ©W.H.Freeman

Harrison S, Blanchet R (2011). Process for preparing highly pure lithium carbonate and other highly pure lithium containing compounds. *US Patent Application 20110200508.*

Hassid M, Ketzinel Z, Volkman Y (1986). Recovery of uranium from wet-process phosphoric acid by liquid-solid ion exchange. US Patent 4,599,221.

Hatch MJ, Dillon JA (1963). Acid retardation: a simple physical method for separation of strong acids from their salts. *Ind Eng Chem Process Design Develop* **2:** 253-263.

Helfferich F (1962). *Ion Exchange.* McGraw-Hill Book Company, Inc.

Hennig C, Schmeide K, Brendler V, Moll H, Tsushima S, Scheinost AC (2007). The Structure of Uranyl Sulfate in Aqueoue Solution-Monodentate versus Bidentate Coordination. *X-ray Absorption Fine Structure-XAFS13, edited by B.Hedman and P.Pianetta, American Institute of Physics, 184-186.*

Hewett JV, Stahl GW (1960). Dimethyl formamide purification. *US Patent 2,942,027*

Hgiem NV, Lee J-C, Jha MK, Jeong J, Hwang TS (2008). Ion exchange of copper from the chloride wash water of electronic industry. In Young CA Hydrometallurgy 2008: Proceedings Sixth International Symposium, 201-208.

Hirosawa K, Kurokawa H, Yamamoto T, Matsunishi M, Nawata Y (1995). Method for purifying organic solution containing lactams. *US Patent 5,440,032.*

Hoffman JG, Dykstra Havlicek M, Yuan W (2003). Intergrated method of preconditioning a resin for hydrogen peroxide purification and purifying hydrogen peroxide. *US Patent 6537516.*

Holger F, Pfeffinger J, Leutner B (2003). Method for producing high purity lithium salts. *US Patent 6,592,832*

Höll WH (1996). Elimination of heavy metals from water by means of weakly basic anion exchange resins. *In* Craig J *IEX'96 Proceedings*, Soc Chem Ind, Hartwells Ltd, Bodmin UK, 404-411.

Honig H, Geigel S (1995). Process for the purification of hydrogen peroxide for microelectronics. *European Patent EP 0502466.*

Horwitz PE, Alexandratos SD, Gatrone RC, Chiarizia R (2002). Phosphonic acid based ion exchange resins. *EP 0618843B1.*

Hsu CKC, Laird DE (1984). Recovery of copper and vanadium from adipic acid production. *European patent ER 0122249 A1.*

Hubicki Z, Wawrzkiewicz M, Wolowicz A (2008). Application of ion exchange methods in recovery of Pd(II) ions-a review. *Chem Anal (Warsaw)* **53:** 759-784.

Hubicki Z, Kołodyńska D (2012). Selective removal of heavy metal ions from waters and waste waters using ion exchange methods. *In* Kilislioglu A *Ion Exchange Technologies*. InTech, 193-240.

Iler RK, Wolter FJ (1951). Silica sol process. *US Patent 2,631,134*.

Iimuro S, Kitamura T, Morimoto Y (1992). Procédé de fabrication de bisphénol A. *European patent EP 0 329 075 B1*.

Inaba Y, Kurokawa Y, Hirakawa T, Oyama K (1993). Ion exchange purification method of aqueous caprolactam solution. *US Patent 5,245,029*.

Ichihashi H, Sato H (2001). The development of new heterogeneous catalytic processes for the production of ε-caprolactam. *Appl Catalysis A: General* **221:** 359-366.

Jeffers TH (1985). Separation and recovery of cobalt from copper leach solutions. *J of Metals* **37:** 47-50.

Jeffrey MI, Anderson CG (2003). A fundamental study of the alkaline sulfide leaching of gold. *European J Mineral Proc Environmental Protection* **3:** 336-343.

Jha MK, Kumar V, Singh RJ (2001). Review of hydrometallurgical recovery of zinc from industrial wastes. *Resources, Conservation and Recycling,* **33:** 1-22.

Jones CP, Pierce AM, Roberts BR (2006). Application of electrochemically regenerated ion exchange to waste water recycle, mixed acid wastes and plating bath rinses. *Paper presented at the Semiconductor Pure Water and Chemical Conference, Feb. 2006.*

Joo SH, Kim YU, Kang JG, Kumar JR, Yoon HS, Parhi PK, Shin SM (2012). Recovery of Rhenium and Molybdenum from Molybdenite Roasting Dust Leaching Solution by Ion Exchange Resins. *Materials Transactions* **53**: 2034-2037.

Jyo A, Kudo S, Zhu X, Yamabe K (2005). Zirconium (II) loaded Diaion CRP200 resin as a specific adsorbent to As(III) and As(V). *Chemistry for the Protection of Environment 4, Vol. 59 of Env Sci Res, p. 29-47.*

Kabay N, Demircioglu M, Ekinci H, Yüksel M, Saglam M, Akcay M, Streat M (1998a). Removal of metal pollutants (Cd(II) and Cr(III)) from phosphoric acid solutions by chelating resins containing phosphonic or diphosphonic groups. *Ind Eng Chem Res* **37**: 2541-2547.

Kabay N, Demircioglu M, Yayli S, Günay E, Yüksel M, Saglam M, Streat M (1998b). Recovery of uranium from phosphoric acid solutions using chelating ion-exchange resins. *Ind Eng Chem Res* **37**: 1983-1990.

Keating JT, Behling KJ (1990). Brine, impurities and membrane chlor-alkali cell performance. *In* Prout NM, Moorhouse JS *Modern Chlor-alkali Technology* SCI, Vol. 4, 125-139

Kelly PP (1984). Process for removing aluminum and silica from alkali metal halide brine solutions. *US Patent 4,450,057*

Kirman LE, Seufer Jr KC (1989). Apparatus and method for recovering materials from process baths. *US Patent 4,863,612*

Koff JL, Zarate DA (1997). Ion exchange removal of cations under chelating/complexing conditions. *US Patent USH1661 H.*

Kolodynska D, Hubicki Z (2012). Investigation of sorption and separation of lanthanides on the ion exchangers of various types.

In Kilislioglu A *Ion Exchange Technologies*, InTech, Chapter 6, 101-153.

Kononova ON, Goryaeva NG, Vorontsova TV, Bulavskaya TA, Kachin SV, Kholmogorov AG (2006). Sorption of thiocyanate silver complexes and determination of silver by diffuse reflectance spectroscopy. *Bull Korean Chem Soc* **27:** 1832-1838

Kononova ON, Goncharova EL, Melnikov AM, Kashirin DM, Kholmogorov AG, Konontsev SG (2010). Ion exchange recovery of Rh(III) from chloride solutions by selective anion exchangers. *Solvent Extraction and Ion Exchange 28: 388-402.*

Kononova ON, Melnikov AM, Borisova TV, Krylov AS (2011). Simultaneous ion exchange recovery of platinum and rhodium from chloride solutions. *Hydrometallurgy **105**, 341-349.*

Koshima H (1986). Adsorption of Iron (III), Gold (III), Gallium (III), Thallium (III) and Antimony (V) on Amberlite XAD and Chelex 100 resins from Hydrochloric acid solution. *Analytical Sciences* **2:** 255-260.

Kotze M, Green B, Mackenzie J, Virnig M (2005). Resin-in-Pulp and Resin-in-Solution. *In* MD Adams *Developments in Mineral Processing, Vol. 15 Advances in Gold Ore Processing Ch. 25, 603-636.*

Kotze MH (2012). What are the major roles of ion exchange in hydrometallurgy? *Presented at the* International Conference on Ion Exchange, SCI, Cambridge, UK, September 18-21, 2012.

Ku Y, Lee KC (2000). Removal of phenols from aqueous solution by XAD-4 resins. *J Hazard Mater* **80:** 59-68.

Ku Y, Wang W (2005). Removal of phenols from aqueous solutions by Purolite A-510 resin. *Separ Sci Technol* **39:** 911-923.

Kumar BN, Radhika S, Reddy BR (2010). Solid-liquid extraction of heavy rare-earths from phosphoric acid solutions using Tulsion CH-96 and T-PAR resins. *Chem Eng J* **160**: 138-144.

Kunin R (1958). *Ion Exchange Resins*. John Wiley, 2nd Edition.

Lanxess (2011a). Product information Lewatit® TP272.

Lanxess (2011b). Product information Lewatit® Monoplus TP214.

Lanxess (2012). Product information Lewatit® VP OC 1026.

Lanxess (2013). Smaller, faster, more stable. Press release April 16, 2013.

Law HH (1983). Recovery of gold in gold plating processes. *US Patent 4,372,830*.

Lee JM, Bauman WC (1983). Removal of sulfate ions from brine using composite of polymeric zirconium hydrous oxide in macroreticular matrix. *US Patent 4,415,677 A*

Leleu R (1983). Procédé de dépolution d'effluents provenant de la production de nitrate d'ammonium et de recuperation des éléments dans ces effluents. *European Patent ER0081121 A1*.

Lillkung K, Aromaa J, Forsen O (2013). Determination of leaching parameters for the recovery of platinum group metals from secondary materials. *Physicochem Probl Miner Process* **49**: 463-472.

Lukey GC, van Deventer JSJ, Schallcross DC (1998). Is ion exchange technology for gold extraction ready for commercialization?. *Aus I Min Metall '98-the Mining Cycle*, Mount Isa, 19-23 April 1998, pp 349-354.

Maketon W (2007). *Treatment of Cu-CMP waste streams containing Cu(II) using polyethyleneimine (PEI)*. PhD thesis, University of Arizona.

Mansoor H (2001). Glycol purification. *EP 1301455 B1*.

Mansoor H (2007). Glycol purification. *European patent application EP 1268377 B1*.

Manziek L (1982a). Borane reducing resins. *US Patent 4,311,811*.

Manziek L (1982b). Amborane® resins: a new approach to the recovery of precious metals. Presented at the 6th International Precious Metals Conference, Newport Beach, CA, June 7-11, 1982.

Maranon E, Fernandez Y, Castrillon L (2005). Ion exchange treatment of rinse water generated in the galvanizing process. *Water Environ Res, 77, 3054-8*.

Maroldo SG, Betz WR, Borenstein N (1990). Carbonaceous adsorbents from pyrolyzed polysulfonated polymers. *US Patent 4,957,897*.

Marsden JO, House CI (2006). *The Chemistry of Gold Extraction*. Society for Mining, Metallurgy and Exploration, Inc. Publishers, 2nd Edition 477-480.

Martins J, Costa C, Loureiro J, Rodrigues A (1984). Recovery of tungsten from hydrometallurgical liquors by ion exchange. *In* Naden D, Streat M *Ion Exchange Technology*. Ellis Horwood Limited, 715-723.

Matejka Z, Eliasek J (1987). Method of separating heavy metals from complex-forming substances of aminocarboxylic acid type, or salts thereof in aqueous solutions. *US Patent 4,664,810*

Matsushita T (1996). Sulfate removal from brine by using amphoteric ion exchange resin. *J Ion Exchange* **7**: *200-208.*

Mecozzi S, West AP Jr, Dougherty DA (1996). Cation-p interactions in simple aromatics: Electrostatics provide a predictive tool. *J Amer Chem Soc* **118**: 2307-8.

Mendes FD, Costa RS, Martins AH (2005). Selective sorption of nickel and cobalt from acid liquors using chelating resins. *In* Mendes FD *Recuperação de Níquel e Cobalto a partir de Polpas de Lixiviação Ácida de Minério Laterítico pelo Emprego de Resinas Poliméricas de Troca Iônica*, Thesis for PhD degree, Chapter 5 (in english), Universidade Federal de Minas Gerais.

Mendes FD (2009). Resin-in-leach process to recover nickel and/or cobalt in ore leaching pulps. *US Application number 20090056500.*

Mikhaylenko M (2011). Purolite® ion exchange resins for recovery and purification of rhenium. *In* German et al *7th International Symposium on Technitium and Rhenium-Science and Utilization, July 4-8, 2011.* Proceedings, Granitsa, Moscow, 2011, 222.

Millar MH, Hardy FRF, Morris GW, Crampton JR (1995). Purification of hydrogen peroxide. *US Patent 5,397,475*

Mina R (1980). Silver recovery from photographic effluents by ion-exchange methods. *J Appl Photographic Eng* **6:** 120-125.

Minz FR, Vajna S (1985). Process for removing sulfates from electrolysis brine. *US Patent 4,556,463 A*.

Mishra H, Yu C, Chen DP, Goddard WA, Dalleska NF, Hoffmann MR, Diallo MS (2012). Branched polymeric media: Boron-chelating resins from hyperbranched polyethyleneimine. *Environ Sci Technol* **46**: 8998-9004.

Monsanto Chemicals (1964). Recovery of copper and vanadium from aqueous streams by ion exchange. *UK Patent GB956403*.

Moore BW (2000). Selective separation of rare earth elements by ion exchange in an iminodiacetic resin. *US Patent 6,093,376*.

Moore SH, Dotson RL (1984). Removal of chlorate from electrolyte cell brine. *European Patent EP 0098500 A1*

Munns K, Sullivan J (1995). Recovery of phosphoric acid used in the bright dip finishing of aluminum. *Eco-Tec Technical Paper 114*.

Naden D, Willey G (1974). Reduction in copper recovery costs using solid ion exchange. *The theory and practice of ion exchange,* An International Conference, Cambridge, UK, July 25-30, 1976. Society of Chemical Industry.

Nakahiro Y, Horio U, Niinae M, Kusaka E, Wakamatsu T (1992). Recovery of gold with ion exchange resin from leaching solution by acidothioureation. *Minerals Eng* **5**: 1389-1400.

Nakamura H, Katoo H, Minejima N, Shimizu H, Satoo A, Nozaki M, Doochi Y (1967). Recovery of iodide ions from anion exchange resins used to extract iodine. *US Patent 3,352,641*.

Naumann AW (1968). Ion Exchange Extraction of Vanadium. *US Patent 3,376,105 A*.

Nesbitt AB, Petersen FW (1994). Recovery of metal cyanides using a fluidized bed of resin. *In* Demirel and Ersayin *Progress in Mineral ProcessingTechnology* Balkema, Rotterdam, 479-486.

Neumann S (2008). IDA resins: versatile specialists. *Speciality Chemicals Magazine, October 2008, 26-27.*

O'Malley GP (2002). Recovery of gold from thiosulfate solutions and pulps with anion-exchange resins. *Thesis submitted to Murdoch University, Perth, W.Australia.*

Oh CG, Ahn JH, Ihm SK (2003). Adsorptive removal of phenolic compounds by using hypercrosslinked polystyrene beads with bimodal pore size distribution. *Reactive and Func Pol* **57:** 103-111.

Oke K, Neumann S, Adams B (2011). Selective fluoride removal. *Water Today* **1:** 76-80.

Owens DL (1980). Chemical Processing by Ion Exchange. *Rohm and Haas Bulletin IE-262.*

Owens M, Goltz RH, Mingee D, Kelly R (2009). Trace Mercury Removal from Flue Gas Desulfurization Wastewater. *IWC*

Pajunen P, Mangum P. Nickel salt recovery using Recoflo® short bed ion exchange. *Eco-Tec Inc., Technical Paper 185.*

Palaniandavar M Velusami M, Mayilmurugan R (2006). Iron (III) complexes of certain tetradentate phenolate ligands as functional models for catechol dioxygenases. *J Chem Sci* **118:** 601-610.

Parschova H, Jureckova K, Mistova E, Jelinek L (2009). Sorption of Germanium (IV) on resin having methyl-amino-glucitol moiety. *Ion Exchange Letters* **2:** 46-49.

Pearson R (1963). Hard and Soft Acids and Bases. *JACS* **85:** 3533-3539.

Pennington L, Williams M (1959). Chelating Ion Exchange Resins. *Ind Eng Chem* **51:** *759-762.*

Pesavento M, Biesuz R, Gallorini M, Profumo A (1993). Sorption mechanism of trace amounts of divalent metal ions on a chelating resin containing iminodiacetate groups. *Anal Chem* **65:** 2522-2527.

Petruzzelli D, Tiravanti G, Passino R (2002). Cr(III) Separation and Recovery from Tannery Wastes : Research, Pilot and Demonstration Scale Investigation. *In* SenGupta AK, *Environmental Separation of Heavy Metals,* CRC Press LLC, 307-349.

Popat KM, Anand PS, Dasare BD (1994). Selective removal of fluoride ions from water by the aluminium form oft he aminomethylphosphonic acid-type ion exchanger. *Reactive Polymers* **23:** 23-32.

Powell JE, Spedding FH, Wheelwright EJ (1957). Method of separating rare earths. *US Patent 2,798,789 A.*

Price K, Novotny C (1979). Water recycling and nickel recovery using ion exchange. *In* Second Conference on Advanced Pollution Control for the Metal Finishing Industry, Kissimmee Fl, USA, Feb. 1979, *Proceedings, 85-87.*

Qureshi N, Hughes S, Maddox IS, Cotta MA (2005). Energy-efficient recovery of butanol from model solutions and fermentation broth by adsorption. *Bioprocess Biosyst Eng* **27:** 215-222.

Rohm and Haas Company (2001). Amberlite IRC748 Product Data Sheet.

Rohm and Haas Company (2006). Ambersep® GT74 Product Data Sheet.

Rumpold R, Antrekowitsch J, (2012). Recycling of platinum group metals from automotive catalysts by an acidic leaching process. *Southern African Institute of Mining and Metallurgy,* Fifth International Platinum Conference, Platinum 2012, Sun City, S.Africa, Sept. 17-21, 2012.

ResinTech (2008). Gold removal and recovery. *Technical Data Sheet 01/08.*

Reynolds BD (1993). Apparatus and process to regenerate a trivalent chromium bath. *US Patent 5,269,905*

Rezkallah A (2012). Method for the recovery of uranium from pregnant liquor solutions. *Patent application US20120125158 A1.*

Rezkallah A (2015). Method for separation of monovalent metals from multivalent metals. US *Patent 9,126,843 B2..*

Riveros PA (2013). Method for removing antimony from copper electrolytes. *US Patent 8,349,187.*

Rohm and Haas Company (1949). Ion exchange without water. *In* Amber-hi-lites N° 3 Ion Exchange Report.

Rohm and Haas Company (2006). Ambersep® GT74 Product Data Sheet.

Rossiter GJ and Carey KC (1998). Copper recovery from leach liquors using continuous ion exchange. *Copper Hydromet*

Roundtable 98, Randol, Vancouver, British Colombia, 17-20 November pp. 185-191

Rossiter GJ, Pease SF, Snyder CB (1992). Process for the removal of vanadium from the wet process phosphoric acid. *US Patent 5,171,548.*

Rourke WJ, Lai WC, Natansohn S (1989). Ion exchange method for the recovery of scandium. *US Patent 4,816,233 A*

Sakaitani H, Sugihara Y, Tanaka K (1997). Process for purification of hydrogen peroxide. *US Patent 5,614,165*

Sassaman, Jr FL, Kaars SR, Filson JL, Kemp PM (2004). Ion exchange removal of metal ions from wastewater. *US Patent 6,818,129*

Schilde U, Uhlemann E (1993). Separation of several oxoanions with a special chelating resin containing methylamino-glucitol groups. *Reactive Polymers* **20**: 181-188.

Schmidt B, Wolters R, Kaplin J, Schneiker T, Lopez-Delgado A, Alguacil FJ (2005). Rinse water regeneration in stainless steel pickling. *Proceedings 9th Int Conf on Envir Sci and Technology, Rhodes, Greece, 1-3 Sept 2005.*

Schmitz PA, Duyversteyn S, Johnson WP, Enloe L, McMullen J (2001). Adsorption of aurocyanide complexes onto carbonaceous matter from preg-robbing Goldstrike ore. *Hydrometallurgy* **61:** 121-135.

Schmuckler G (1969). Method for recovery of noble metals. *US Patent 3,473,921.*

Schoeman E, Bradshaw SM, Akdogan G, Eksteen JJ (2013). The recovery of platinum, palladium and gold from cyanide heap solution with the use of ion exchange resins. In Glen HW A catalyst for change, the 5th International Platinum Conference, Sep 18-20, 2012, pp 729-742. Sun City, South Africa, The South African Institute of Mining and Metallurgy.

SenGupta AK, Cumbal LH (2007). Hybrid anion exchanger for selective removal of contaminating ligand from fluids and methods of manufacturing thereof. *US Patent 7,291,578.*

SenGupta AK, Padungthon S (2013). Hybrid anion exchanger impregnated with hydrated zirconium oxide for selective removal of contaminating ligand and methods of manufacturing and use thereof. *US Patent 20130274357 A1.*

Soldatov VS, Kosandrovich EH (2011). Ion exchangers for air purification. *In* Sengupta AK *Ion Exchange and Solvent Extraction, a Series of Advances,* Vol. 20, 45-110.

Sricharoenchaikit P (1989). Ion Exchange Treatment for Electroless Copper-EDTA Rinse Waters. *Plating and Surface Finishing, December 1898, 68-70.*

Strong B, Henry RP (1976). The purification of cobalt advance electrolyte using ion exchange. *Hydrometallurgy* **1:** 311-317.

Suppanen IJP (2004). Experiences of implementing a grass root investment project in Aqaba, Jordan. *Presented at the IFA Technical Conference, Beijing, China,* April 20-23, 2004.

Tadao N, Yashiaki E (1985). Method for purification of sulfuric acid solution. *US Patent 4,559,216.*

Tatarnikov AV, Soloskaya I, Shneerson YM, Lapin AY, Goncharov PM (2004). Treatment of platinum flotation products. *Platinum Metals Review* **48**: 125-132.

Taut JJ, Sole KC, Hardwick E (2013). Zinc removal from a base metal solution by ion exchange: process design to full-scale operation. *S.African Inst Mining and Metallurgy, Base Metals Conference, 2013.*

Ten Kate A, Manuhutu CFH, Bargeman G, Bakkenes H, Westerink JB, Steensma M, Demmer RLM (2009). Nanofiltration as energy-efficient solution for sulfate waste in vacuum salt production. *Desalination* **246**: 87-95.

Tidwell PW (1957). Methanol purification. *US Patent 2,792,344.*

Tjioe TT, Weij P, van Rosmalen GM (1987). Removal of cadmium by anion exchange in a wet phosphoric acid process. *In* KJA De Waal et al (Editors) *Environmantal Technology* Martinus Nijhoff Publishers, Dirdrecht, 1987, p. 145-147.

Tongesayi T (2014). Method for removing contaminants from water using thiol-modified surfaces containing ester linkages. *US Patent 8,828,235 B2.*

Trémillon B (1965). *Les separations par les resins échangeuses d'ions.* Gauthier-Villars Paris.

Trochimczuk AW, Jezierska J (2000). New amphoteric/ion exchange resins with substituted carbamylethylenephosphonates; synthesis and EPR studies of their Cu(II) complexes. *Polymer* **41**: 3463-3470.

Troshkina ID (2011). Rhenium in nuclear fuel cycle. *In* German et al 7th *International Symposium on Technitium and Rhenium-*

Science and Utilization, July 4-8, 2011. Proceedings, Granitsa, Moscow, 2011, 202-207.

Tverdislov VA, Yakovenko LV, Salov DV, Tverdislova IL, Hianik T (1999). The parametric pump mechanism in separation of components in heterogeneous systems. I. Macroscopic distributed systems. *Gen Physiol Biophys* **18:** 73-85.

van Deventer J, Wyethe JP, Kotze MH, Shannon J (2000). Comparison of resin-in-solution and carbon-in-solution for the recovery of gold from clarified solutions. *J South African Inst Mining Metallurgy,* July/August, 221-227.

van Deventer J, Bazhko V, Yahorava V (2014). Comparison of gold-selective ion Exchange resins and activated carbon for the recovery of gold from copper-gold leach liquors. *Proceedings ALTA 2014 Gold-Precious metals sessions, Perth, Australia.*

Virolainen S (2013). Hydrometallurgical recovery of valuable metals from secondary raw materials. *Thesis for PhD, University
of Lappeenranta, Finland.*

Volkman Y (1987). Recovery of uranium from phosphoric acid by ion exchange. *In* The recovery of uranium from phosphoric acid. *IAEA Tecdoc-533, IAEA, Vienna 59-68.*

Wall J (2009). Comparison of methods for the purification of biodiesel. MSc Thesis, University of Idaho.

Wan RY, Levier KM (2008). Precious metal recovery using thiocyanete lixiviant. *EP 1629129 A4*

Weiner R (1963). *Effluent treatment in the metal finishing industry.* Robert Draper Ttd, Teddington, 167-168.

White IF, O'Brien TF (1990). Secondary Brine Treatment : Ion Exchange purification of brine. *In*: Prout NM, Moorhouse JS *Modern Chlor-alkali Technology.* SCI, Vol. 4, 271-289

Winget OJ (1971). Separation of rare earth elements by ion exchange. *US Patent 3,615,173.*

Woyski MM (1965). Ion exchange process for separating the rare earths. *US Patent 3,167,389.*

Yahorava V, Kotze MH (2011). RIP pilot plant for the recovery of copper and cobalt from tailings. *The Southern African Institute of Mining and Metallurgy 6th Southern African Base Metals Conference.*
Yamaguchi K (1990). Effects of mercury in brine on the performance of the membrane-type chlor-alkali plant. *J Electrochem Soc* **137** : 1423-1430.

Yoshida A (2006). Silica nucleation, polymerization and growth preparation of monodispersed sols. *In* Bergna HE, Roberts WO *Colloidal Silica Fundamentals and Applications.* CRC Press Taylor and Francis.

Zaganiaris EJ (2009). *Ion Exchange Resins in Uranium Hydrometallurgy.* Books on Demand.

Zaganiaris EJ (2011). *Ion Exchange Resins and Synthetic Adsorbents in Food Processing.* Books on Demand.

Zainol Z (2005). The development of a resin-in-pulp peocess for the recovery of nickel and cobalt from laterite leach slurries. *PhD thesis, School of Mineral Science, Murdoch University, Murdoch, Western Australia, Australia.*

Zhao Z, Li X, Zhao Q (2010). Recovery of V2O5 from Bayer liquor by ion exchange. *Rare Metals* **29:** *115-120*

Zipperian DC, Raghavan S (1985). The recovery of vanadium from dilute acid sulfate solutions by resin ion exchange. *Hydrometallurgy 13: 265-281.*

Subject Index

A

Acid mine drainage (AMD)   180-182
acid retardation   70-72, 100, 104, 105, 123, 135
adipic acid   137
aldehydes removal   211
aluminium anodizing   104-107
aluminium removal
    from brines   196
    from H_2O_2 207
    from H_3PO_4 104
Amberlite® IRC50   249, 254
Amberlite® PWA5   160
Amberlite® XAD4   62, 224
Amberlite® XAD7   35, 154, 159
Ambersep® GT74   53, 234, 255
Ambersorb®   245
Amborane®   61, 159
amidoxime   58, 180
aminomethylphosphonic resins   37, 44-45, 190, 219
ammonium removal from cuprammonium spinning   139
ammonium nitrate recovery   240
antimony   167, 237
arsenic   174, 237-238

B

Barium removal from brines 188, 189
Base metals 61, 130, 141, 150, 151, 160, 161-167
Biodiesel purification 256
Bismuth 167
Bis-picolylamine (BPA) 48-52, 102, 119, 121, 127, 128, 134, 136, 162, 164
Boron hydride 61, 158, 159
Boron removal 34, 59, 202-205
Brine purification 185-201

C

Cadmium 95, 110, 124, 181, 219, 240
Caprolactam purification 212-215
Cation-π interaction 35
Caustic purification 221
Chemical mechanical polishing (CMP) 92, 119, 151
Chlorates 187, 201
Chromic acid recovery from Al anodizing 106-107
Chromium plating 111
    Chromate bath recovery 112-113
    Rinse waters recycling 114-118
    Cr^{3+} plating 118
chromium (III) removal 41, 112, 115, 133, 162, 219
Cobalt 47, 48, 152, 161, 162, 166, 181, 235
Cobalt electrolyte 51, 164
colloidal silica, see: silica sols
Copper
    from adipic acid 137

    from AMD   181
    from Cr^{3+} baths   119
    from cobalt electrolyte   163
    from cuprammonium spinning effluents   139
    from electroless plating   127
    from leach liquors   165
    from rinse waters   119-122, 123
Cyanides removal from wastes   155-158
Cyanides, gold   142-155
Cyanides, leach solutions, PGM   160
Cyanides, rinse waters   117

D

D2EHPA   62, 63, 103, 136, 163, 165, 178, 220
Diaion® CR-11   44
Diaion® CRP200   48, 238
Diaion® DSR01   75, 200
Diaphragm cell   186
Dicalcium phosphate   228
Dimethyl formamide purification   209
Diphonix® resins   47, 103, 136, 172, 178, 182, 219, 221, 238
Donnan equilibrium   36, 64-69, 70, 97, 237
Dowex® M-4195   50-51, 119, 162, 163, 164
Dowex® Retardion® 11A8   74, 200, 221, 222
Dowex® XFS-43084   51, 165, 166
Dowex® XUS-40323   217
Dowex® XUS-43436   217
Dowex® XUS-43578   50
Dowex® XUS-43605   51
Dowex® XZ-91419   148, 151

E

EDTA 29-31, 32, 43, 50, 111, 127, 128, 134, 137, 176, 178, 179, 221, 236
Electrodeionization 77
Electroless plating 110
Electroplating 107
Ethylenediamine 27, 31, 122

F

Fertilizers 225
Fluorides 100, 112, 236
Formaldehyde purification 215

G

Gallium 58, 179
Galvanization 124-127
Germanium 61
Glycols purification 210
Glycerin 70, 256, 257
Gold
    Hydrometallurgy 142-155
    Recovery from plating baths 128-130

H

Hafnium 168
Hard and Soft Acids and Bases (HSAB) 26
Heavy metals removal from wastes 233
Hydrochloric acid purification 216

Hydrogen peroxide purification 205
Hydrometallurgy 141
Hydroponics 229
Hydroxypropylpicolylamine (HPPA) 49-51, 127, 128, 134, 166, 167

I

Iminodiacetic acid (IDA) 27-28
Iminodiacetic acid resins 38-44, 119, 120, 125, 127, 136, 163, 178, 181, 188,
Incinerators wastes 238
Indium 110
Iodides removal from brines 197-200
Ionac® SR 4 53
Ion exchange equilibrium 79
    favorable 81
    unfavorable 81
    mono-divalent 85-88
Ion exclusion 69
Ion retardation 73
Iron
    from Cr^{3+} solutions 118, 119
    from galvanization solutions 124-126
    from H_2SO_4 pickling liquors 99
    from HCl purification 216

L

Lead 47, 233-234, 237, 240
Lewatit® TP 207 44, 163
Lewatit® TP 214 54
Lewatit® TP 220 42

Lewatit® TP 272　　　　　63
Lewatit® VP OC 1026　　　62, 102, 163, 178
Ligand exchange　　　　　35, 36, 238
Lithium　　　　　　　　　201-202

M

Magnesium chloride brines, boron removal　　　202
Manganese　　　48, 102, 164, 176, 182, 235
Mercury　　　　32, 33, 52-58
    from brines　　　　　　　195
    from hydrocarbons　　　　255
    from incinerators wastes　240
    from wastes　　　　　　　234
Methanol purification　　　　　　　　254
Methylglucamine　　　　　　　　　　59
Minix®　　　　　150, 160
Molecular Recognition Technology　　　61
Molybdenum　　　141, 169-171, 174, 175, 237

N

Nickel removal　　47, 50, 63, 94, 110, 116, 135, 181
    from brines　　　　　　　　197
    from cobalt electrolyte　　　　51, 164
    from electroless plating　　　　127
    from lateritic ores　　　　　　161-162
    from plating rinse waters　　　122-123
Niobium　　　　　　169
Non-aqueous systems　　　　　249

O

Organics removal 138, 145, 154, 205-206, 240, 245
    From plating rinse waters 132
    from wastes 241

P

Palladium 56, 110
Parametric pumping 20, 75
Passivation 92, 102-103, 136
Phenol 34, 235
Phenol deacidification 252-253
Phenols removal from wastes 241-242
Phosphinic resins 33, 46, 48, 63, 178, 179, 220
Phosphonic resins 33, 44, 46-47, 219, 238
Phosphoric acid purification 179, 218-221
    bright dip finishing 104
    passivating baths 102-103
Photographic baths 223
Pickling liquors
    HCl 95
    H_2SO_4 99
    Mixed HF/HNO_3 acids 99
Picolylamine, see bis-picolylamine
Platinum group metals
    Hydrometallurgy 158
    Recovery from plating baths 130
Polyamine resins 56-57
Polyethylene imine (PEI) resins 57, 61
Potassium chloride brines 202
Preg-robbing 150, 154
Purolite® A-170 172, 173, 174
Purolite® A-172 172, 173, 174
Purolite® S-957 49, 103, 120

Purolite® S-960	52
Purolite® S-984	236
Pyridine	27, 50, 59, 61, 131, 132, 161, 255, 256

R

Radioactive wastes	48
Rare earths	137, 142, 175-179, 221
Recoflo®	72, 73, 75, 101, 105
Reillex® HP425	160
Rhenium	171-174
Rhodium	110, 131
Rinse waters (plating)	
Acid rinse waters in chromate plant	115
Alkaline rinse waters	117
Cadmium	124
Copper	121-123
Nickel	129
Organics	133

S

SAD (strong-acid dissociables)	117, 155, 157
Scandium	176, 177, 222
Selectivity coefficient	79, 80, 84, 85, 89
Selectivity sequence for SAC resins	181
Separation factor	79, 80, 81, 84
Silica removal from brines	196
Silica sols	229-232
Solvent impregnated resins (SIR)	49, 62-63, 164, 181

Sulfuric acid recovery from Al anodizing baths 104

T

Tantalum 168
Thermal regeneration 76
Thallium 236
Thiol resins 53-54, 137, 195-196, 234, 236, 237, 238, 242, 257
Thiourea resins 54
Thiouronium resins 55
Thorium 178, 182
Tulsion® CH-96 48, 221
Tungsten 142, 176, 177

U

Uranium 47, 59, 60, 142, 154, 163, 166, 169, 170, 172, 174, 175, 176
Uranium from phosphoric acid 221

V

Vanadium 138, 142, 174-175, 220, 238
    from adipic acid 137
volatile organic compounds (VOC) 246-247

W

WAD (weak-acid dissociables) 117, 147, 155, 157

Z

Zinc
- From Cr^{3+} solutions — 119
- from cobalt electrolyte — 165
- from geothermal brine — 182
- from leach liquors — 166
- from HCl pickling liquors — 95
- from viscose rayon — 138-139

Zirconium — 168, 238